DARK SIDE OF THE OCEAN

THE DESTRUCTION OF OUR SEAS, WHY IT MATTERS, AND WHAT WE CAN DO ABOUT IT

Albert Bates

GroundSwell Books
Summertown, Tennessee

Library of Congress Cataloging-in-Publication Data

Names: Bates, Albert K., 1947- author.

Title: Dark side of the ocean : the destruction of our seas, why it matters, and what we can do about it / Albert Bates.

Description: Summertown, Tennessee : GroundSwell Books, [2020] | Includes bibliographical references and index. | Summary: "Ocean biodiversity is being decimated on par with the fastest rates of rain forest destruction. More than 80 percent of pollutants in the oceans come from sewage and other land-based runoff (some of it radioactive). The rest is created by waste dumped by commercial and recreational vessels. In many areas and for many fish stocks, there are no conservation or management measures existing or even planned. Climate author Albert Bates explains how ocean life maintains adequate oxygen levels, prevents erosion from storms, and sustains a vital food source that factory-fishing operations cannot match—and why that should matter to all of us, whether we live near the ocean or not. He presents solutions for changing the human impact on marine reserves, improving ocean permaculture, and putting the brakes on the ocean heat waves that destroy sea life and imperil human habitation at the ocean's edge"— Provided by publisher.

Identifiers: LCCN 2020022097 (print) | LCCN 2020022098 (ebook) | ISBN 9781570673948 (paperback) | ISBN 9781570678271 (epub)

Subjects: LCSH: Marine biodiversity conservation. | Marine pollution—Environmental aspects. | Endangered ecosystems. | Marine ecology.

Classification: LCC QH91.8.B6 B38 2020 (print) | LCC QH91.8.B6 (ebook) | DDC 333.95/616—dc23

LC record available at https://lccn.loc.gov/2020022097

LC ebook record available at https://lccn.loc.gov/2020022098

We chose to print this title on sustainably harvested paper stock certified by the Forest Stewardship Council, an independent auditor of responsible forestry practices. For more information, visit us.fsc.org.

MIX
Paper from responsible sources
FSC® C005010
www.fsc.org

Printed in the United States of America

GroundSwell Books
an imprint of BPC
PO Box 99
Summertown, TN 38483
888-260-8458
bookpubco.com

ISBN: 978-1-57067-394-8

25 24 23 22 21 20   1 2 3 4 5 6 7 8 9

CONTENTS

Introduction

I am fortunate these days to find myself in a place where I can swim every day. I head offshore about 250 yards and then paddle along, parallel to the beach for a quarter mile, before returning to where I started. Tucked inside the great Mesoamerican Reef, the shallow turquoise waters in this part of the world are usually safe from sharks and jellyfish. The waves are calmer, making swimming easier on this old body.

We are of the oceans, you and I. Floating in the sea, gazing up into a blue sky, I return to human origins. My cells are from the single-cell life-forms of the sea. Indeed, the amniotic fluid I "breathed" for my first nine months in 1946 had about 2 percent salinity, only about a third less than the ocean.

While we speak reverently of *Madre Tierra* and *Terra Firma*, all life depends on water, the *élan vital*, the universal solvent, *aqua mater*. Our mythology is full of stories of great deliverance—the raising of Mount Ararat out of floodwaters to heel Noah's Ark, the landing of the Pilgrims at Plymouth Rock, George Freeth's heroic rescue of eleven Japanese fishermen caught in a gale off Venice Pier in 1908. In a small fishing village in Japan, they still light candles to remember that deliverance and rebirth.

In the course of this book, I have to reference some significant numbers and long periods. Most of us can't relate when we speak of a billion or trillion of something. We can't wrap our heads around something that big. Also, if I say something could happen in 2050 or 2100, that is difficult to picture because we don't know what the world will look like in 30 years or 70 years. If we look back 30 years, to 1990, the world may not seem very different—but no one had smartphones, few even had email or text messaging, and the World Wide Web did not exist. There were no digital cameras, CDs or DVDs, hybrid cars, GPS, or genetically modified foods. If we go back farther, to 70 years, most households did not have a TV; offices did not have copiers, computers, or coffee makers; and there were no commercial passenger jets, nuclear plants, or a NASA space program. So, how can we possibly imagine ourselves in the year 2100?

To get past that difficulty, whenever possible, I am going to use the example of an eight-year-old boy named Xen. He was born in 2012 and will be 18 in 2030, 38 in 2050, and 78 in 2100. In his lifetime, he will witness a world that changes more each year than it would have changed in decades for his grandparents. The world that exists for Xen today is still the world of his parents. It is what it is, and he has only just arrived.

This book will explore how the oceans are today and how that may change during Xen's lifetime and beyond. We begin with the origins of life, move on to the arrival of early humans and their changing relationship with the sea, and then peer into the depths to see what changes that has caused and what is likely to happen in the future. We look at temperature, oxygen, acidity, ice, sea level, ecosystems, populations, and, finally, how all this affects humans on land. It is a fascinating journey, but one that may also be somewhat alarming, as it should be. Over the past thousand years, humans have set in motion some very unexpected trends. Those trends are now starting to bite us. It is not too late to stop the worst of it if we can act quickly enough, but it will be up to Xen's generation to make those decisions. The time remaining is very short.

As I swim along looking at the coastline, I see massive new hotels and homes of concrete and steel, the materials brought in by barge, literally weighing down the sandy beach, more each day—the sand and gravel from quarries in Yucatán, the steel from China, the cement from factories in distant cities. From where I sleep, I can hear the first barge unload even before the first chirps of the dawn chorus.

Each year, more than 92 gigatons (1 Gt is a billion long tons) of these materials—metals, minerals, fossil fuels, and biomass (mostly food)—are drawn out of the Earth and deposited in coastal places like this. This number is growing at a rate of 3.2 percent per year, doubling roughly every 20 years.

Since 1970, extraction of fossil fuels has increased from 6 Gt to 15 Gt a year; minerals such as sand and gravel for concrete have gone from 9 Gt to 44 Gt; biomass harvested for energy and construction has risen from 9 Gt to 24 Gt, and that production is accelerating now.

Land-use change for agriculture, subdivisions, and mountaintop mining accounts for over 80 percent of biodiversity loss and 85 percent of stress on drinking water supply, even without factoring in the fertilizers, pesticides, and herbicides. The entire extraction economy accounts for 53 percent of climate change, even before the fuels are burned. What will it be at the next doubling, expected around 2040, when Xen is 28? Without change, resource demand will more than double to 190 Gt per year, greenhouse gases will rise 40 percent, and demand for land will increase by 20 percent.

Since 2000, we have witnessed what is being called the "blue acceleration." Frenetic economic expansion has placed unprecedented human pressure on marine ecosystems, which are experiencing overfishing, acidification, heat waves, plastics, noise pollution, and radioactivity. But, the rhetoric of a "blue economy," which would combine economic growth with sustainable use, is placing unwarranted expectations on antiquated, complex, and uncertain governance structures.

Jonathan Watts, writing in *The Guardian,* observes that for rich countries, our resource extraction works out to the weight of two elephants per person per year. In emerging countries, it is about the weight of two giraffes, appearing less as second homes on a ski slope or beach and more as smaller items like mobile phones. The piles of materials that went into making these objects are invisible to the consumer.

The UN's Sustainable Development Goal #8 calls for "sustained and inclusive economic growth," which it proposes to accomplish by expanding access to financial credit to create more jobs. Looking at this from a biological perspective, in any ecosystem—and human-dominated ecosystems are no exception—individuals and institutions can only exist at a scale determined by biophysical limits. Whenever limits are uncertain, it is better to err on the side of caution than to suffer collapse. In any system, production and consumption are a balanced pair. Overproduction without consumption creates waste and pollution. Overconsumption without compensating production sets up a system crash.

For more than the past half century, *ecological* economists like Herman E. Daly, Manfred Max-Neef, Hazel Henderson, and Robert Costanza have proposed a useful substitution for *both* production and consumption. Instead of viewing higher-intensity consumption or successful shopping trips as appropriate societal goals, these economists suggest that after meeting basic human needs we could focus on supplying subjective well-being, which will not be found in ever-greater material extraction, but instead in the pursuit of enjoyable lives that do not tax the planet's regenerative capacity.

That kind of economy would be much better for the health of the ocean too. But whatever we do, something needs to change, and soon.

The ocean has absorbed more than 90 percent of the heat humans have added to the Earth's system since the Industrial Revolution. The Arctic is poised for "blue ocean," ice-free summers sometime in the next decade. Even if, through a combination of sharp emissions reductions, biochar, and reforestation, we were to bring greenhouse gases down to where they were 200 years ago, it would take thousands of years for the ice in Antarctica to form again. In the meantime, a combination of all that melted polar ice and the thermal expansion of water means that the 1.9 billion people who live on or near a coastline (about 28 percent of the world's population) may have to relocate.

The ocean plays a significant role in regulating climate and weather. As we shall see later in this book, that means we can anticipate increased marine heat waves, coastal flooding, extreme tropical events, and wildfires. What happened in Houston in 2018 and Australia in 2019 through 2020 is only a preview.

Changes in ecosystems, such as ocean oxygen, acidification, and sea ice loss, have caused ocean animals to migrate in search of new food sources. Local and global extinctions are cascading. Because of overfishing and pollution, the ocean's health as a living organism is in peril. That will have a dollar cost, currently put at $428 billion per year by 2050 and $1.98 trillion by 2100. When so many species suffer, the effects ripple out to the entire biosphere, affecting us all.

Adaptation and conservation measures will determine survival for many species. Individuals, organizations, and governments need to increase efforts to build resilience. Politicians need to become leaders in cutting carbon emissions and protecting their local constituents with better environmental regulations.

There are many things we can do. Blue carbon initiatives prohibiting deep-sea mining, replanting mangroves, protecting salt marshes and seagrasses, curtailing cruise ships and trawlers, and restoring coral reefs give multiple co-benefits. We know how to do these things, and we know it is more cost-effective and financially rewarding to do them soon rather than wait. Let's get going.

PART ONE

CHAPTER

1

We Are the Salt of the Sea

If you were to wake up one morning and all you saw around you was a vast expanse of open ocean, that would be an average landscape of the Earth. Our tiny planet, orbiting around its sun and slowly rotating, shines blue against the blackness of space because of all this water.

Seven-tenths of Earth's surface is water, and 97 percent of that is in the ocean. Before we knew any better, we gave the parts of this big wet thing names like Atlantic, Pacific, Indian, Arctic, and Southern, but those are just descriptions of its regions because it is all one single ocean. At its deepest, it is deeper than Mount Everest is tall. At its broadest, it crosses 13 time zones from beach to beach.

If we removed the seas and walked down into the dry basin, we would be, on average, about a mile deeper than the beach where we started. The bottom would not be a bowl, however, but a varied terrain of mountains, valleys, canyons, and giant boulders.

When you put back the water, you immediately notice it is not still but has strong currents moving one way or another: sideways, up, and down. Those currents have been so strong in some places and at some times that they have rolled a few of those giant boulders up onto a beach and left them there to dry.

Let's use the example of our young man named Xen again, because we need to talk about the size and age of the world in relation to mere humans.

Long before there could even be an eight-year-old boy, more than four billion years ago, when the Earth formed, you could think of it as a spongy rock. The sponginess was because it was not solid, but moist—nearly all its water embedded in pores in the red-hot rock. As it cooled, a process that took at least a billion years (so far, because it is still going on), a lot of that water condensed into a thin skin covering the rock's cooler external surface. You can think of it like the skin on an apple. Even though Earth's ocean can be up to four miles deep in places, our ocean and watery atmosphere are still just a thin skin compared to the size of the planet. We call our planet's skin the biosphere.

About half a billion years after the rock formed, the first microscopic seeds of life began, somewhere in that condensing biosphere. They likely began as something we would today call a bacterium, which could feed itself and reproduce by breaking apart CO_2 into separate molecules of oxygen and carbon, using the energy of sunlight.

Today, with the latest genetic sequencing devices and DNA libraries, we can trace the origins of modern bacteria to these ancient ancestors and then to the formation of the first protein strings. The genetic sequence, as it evolved by trial and error, formed a record, a kind of clock, that we can run backward to learn when different proteins took shape and how organisms eventually came to look the way they do today. Gradually, proteins fit together, absorbed and shed oxygen and nitrogen, and added carbon to themselves to grow more of themselves, which is to say they "ate," inhaled, and exhaled, so much so that over the next few billion years they filled the atmosphere with compounds of oxygen and nitrogen and created the ideal conditions for more complicated forms of life, such as Xen, you, and me, to exist. Other green and purple bacteria came along to eat, exhale, or excrete various combinations of hydrogen, sulfur, iron, calcium, sodium, molybdenum, and other nutrients, and in the process created webs of differentiated single-celled organisms that set the stage for the next great leap in evolution—multicellular organisms.

Beginning about 542 million years ago, and over the next 20 million years, there was a biological explosion—a proliferation of thousands upon thousands of new and different forms of life. Many of the ancestral creatures went extinct, although some of the world's oldest—lobsters, crabs, millipedes, spiders, and some flying insects—still survive. Plants also evolved rapidly and covered the land with ferns, mosses, conifers, and flowering, seed-bearing varieties of all shapes and colors. As their roots extended into the ground, they formed beneficial relationships with the soil microbes that converted minerals into food and released nitrogen, phosphorus, potassium, and other elements that found their way back to the ocean, fertilized the marine food chain, and made it possible for the evolution of larger animals, like swimming dinosaurs and giant sea turtles. As recently as five million years ago, sharks the size of whales roamed the seas, with teeth the size of little Xen.

Five mass extinctions caused the vast majority of species that evolved to disappear, and the worst of those events, so far, came 251 million years ago at the end of the Permian Era, when more than 90 percent of ocean life and two-thirds of land species disappeared from the fossil record. That extinction event may have begun with a half-million-year outbreak of volcanoes. As the lava flooded an area of 620,000 square miles in what is today's Siberia, it burned away peat and coal deposits and released enormous volumes of carbon dioxide, which over thousands of years warmed the Earth 6°C (11°F). The ice at the poles melted, and the oceans rose. Vast deposits of methane frozen undersea or in permafrost warmed and bubbled up into the atmosphere, causing a runaway warming that fed upon itself. Land plants withered, and deserts formed. The ocean became acidic from all the CO_2 and methane it had to absorb. Corals, sea urchins, sponges, and shelled animals died. While this event nearly wiped life off Earth entirely, it eventually subsided short of that, and over the next 10 million years, corals revived; sea urchins, sharks, and turtles returned; and new species filled the oceans and spread across the land.

Our ancestors began in aquatic environments and, like salamanders and mosquitoes, passed from something like gill breathing to air breathing. When animals emerged from the ocean to live on land, they needed lungs—to take oxygen into their blood and exhale the wastes of cellular metabolism to keep the body's elements in balance.

Ecosystems have a balance to maintain, as well. They have stomachs and lungs much as we do and also have to eliminate, or at least recycle, their wastes. Throughout evolution, these systems moved from very simple, growth-oriented, trial-and-error functioning to more sophisticated, steady-state relationships of very diverse and complex groupings of plants, animals, and microbes.

Keystone species are those upon which entire ecosystems rely. While every species is important to its ecosystem, if you take a keystone species away, the system can crumble. Marine animals form relationships with each other that help define their roles. Two organisms might have a predator/prey relationship, like a sea otter and a sea urchin, or they might have a friendlier sort of relationship called mutualism, like when a cleaner wrasse cleans a moray eel's mouth by snacking on the stuff between the eel's teeth.

When we evolved from reptiles to land dwellers, we needed to develop a colon to retain and conserve our internal body fluids. The colon gave us time to remove excess water from digestive wastes before expelling them. Our more transparent marine skins needed to adapt to shield the body from stronger solar radiation, especially ultraviolet, and to better regulate heat, using hair follicles and sweat glands. Inside our organs and glands, however, are a facsimile of the microbial communities we evolved with in the ocean. Each of us has millions of tiny microbes living within and between our cells to help us digest our food, transfer oxygen to our blood, move electric signals to our brains, and communicate with the microbial world outside.

Our blood is salty like the sea for a good reason. Blood is a private ocean that mirrors what life was like for microscopic, single-celled organisms that dominated the first billions of years of life on Earth. The watery portion of blood, the plasma, has a concentration of salt and other ions that is remarkably similar to seawater.

Seawater is the circulatory system of our planet. When Earth gets a fever, her blood runs hot, and she responds by perspiring (casting off more heat to space and making more rain), taking deep breaths (more absorbed carbon dioxide and fiercer winds), and drinking more water (melting ice, floods, and super-typhoons).

When we and other animals left the ocean, we chose to carry water with us as an internal store since we could no longer be sure of a continuous supply. Our skins converted our bodies into portable canteens. Today, over 70 percent of us is water, and the lymph system is the internal ocean we carry about with us. All our organs float in this sea of fluid; our intracellular pericardial blood, cerebral, and spinal fluids are fed by electrolytes regulated by our kidneys. Our respiratory tracts—nostrils, sinuses, trachea, bronchi, and lungs—and digestive and reproductive systems are lined with another saltwatery ring of protection, our mucous membranes.

Some creatures returned to the ocean some 50 million years ago after having already evolved to mammals resembling dogs or cats. Earlier in their evolution, without water to provide buoyancy, these animals had lined their skeletal joints—including between vertebrae—with synovial tissue to pad and lubricate joints against the greater force of gravity. Once freed of the gravity of land, their pelvises reduced in size and separated as their vertebral columns extended to improve locomotion. Because of this, dolphins and whales swim with horizontal tail fins that move up and down, rather than back and forth like the vertical tail fins of fishes. Their backbones bend up and down like those of dogs or cats when they are running.

Humans share with elephants, iguanas, turtles, marine crocodiles, sea snakes, seals, and sea otters our ability to weep salt tears. Of the various salts found in solution in our bodily fluids and the ocean, by far the commonest is table salt, or sodium chloride ($NaCl$). On average, there is slightly more than 6 milligrams of $NaCl$ dissolved in each milliliter of our tears' lacrimal fluid. Average ocean salinity is 3.5 percent, or nearly six times our tears (35 mg/ml). Most of that salt is the same as in our bodies—sodium chloride.

More than 80 percent of ocean pollution comes from land-based activities. As chemicals from agriculture and other human activities wash from land, they increase the ocean salinity. The saltier a body of water is, the less likely it is to absorb carbon dioxide from the atmosphere, and the more likely it is to give it off. This absorptive capacity is a crucial recovery element at the end of ice ages, when salinity peaks due to ice impoundment of fresh water from rain or snow, causing more CO_2 to off-gas to the atmosphere and to positively force the greenhouse effect, rewarming the world. It is not, however, an effect we want to be encouraging when we are trying to reduce global warming.

From coral bleaching to sea-level rise, entire marine ecosystems are rapidly changing. We can see many of these changes in newspaper headlines and scientific reports:

- Many pesticides, fertilizers, and animal pharmaceuticals end up in rivers, coastal waters, and the ocean, resulting in oxygen depletion and toxins that kill or maim marine plants and shellfish.

- Factories and industrial plants discharge sewage and other runoff into the oceans. This, too, results in oxygen depletion and toxins that kill marine plants and shellfish. In the US, sewage treatment plants discharge twice as much oil each year as tanker spills or drilling disasters.

- Oil spills like Deepwater Horizon and nuclear spills like Fukushima pollute the oceans, although air pollution is responsible for almost one-third of the toxic contaminants entering the water.

- Microplastics—dumped by cruise ships, spilled from container ships, or making their way via land or river from factories and garbage dumps—will soon outweigh all the fish in the sea.

- Countless plants and animals, including invasive species such as poisonous algae and cholera, have crossed the border between land and harbor waters and disrupted the ecological balance, due to transoceanic commerce and other human activities.

- Many kinds of seafood are either fished to capacity or overharvested. As the climate rapidly changes and microplastics take their toll, the ability of fish to replenish their populations is dropping dramatically, leading to fishery exhaustion, fish extinctions, and wider famines, seen and unseen.

The ocean is not merely our birth home; it sustains us now. It is possible to live within its limits and the limits of the good Earth. The sooner we can learn how to do that, and get on with it, the better off we will be.

CHAPTER

2

The Great
Blue Highway

As early humans set out from Africa hundreds of thousands of years
ago to colonize the world, they sometimes had to cross rivers. Those
water journeys prepared them for short sea journeys, such as crossing
the Red Sea near Aden or the Persian Gulf near Hormuz. Early boat-
men might have made canoes of bamboo, log, or bark or rowboats of
woven reed mats to get across.

At roughly the same time as the Vikings left their coastal settlements on fjords in Norway and explored the coastlines of England
and France and St. Brendan crossed the Atlantic to Newfoundland,
Hawaiians were discovering New Zealand in the Pacific. The boats
they built were very different in these distant places. The Vikings liked
long, sleek, wooden faerings and knarrs, with elegantly carved bow-
sprits that glided through the water on the muscles of strong oarsmen.
The Polynesians preferred catamarans, trimarans, and outrigger
canoes with small sails. St. Brendan paddled his way from the Kerry
coast of Ireland to Iceland and back in what looked like a floating hot
tub: a round basket of wattle covered in hides tanned in oak bark and
softened with butter.

Of course, over time, boatbuilding craft improved, and as different cultures began to encounter each other, they picked up important design points. Canoes got bigger, with different shapes of sail (lateen, square, claw, and the upside-down triangles known as sprit sails) and arrays of square-rigged or junk-rigged sails. More substantial and heavier hulls were able to carry men, cattle, horses, and cargo over enormous distances. A nineteenth-century whaler might stay at sea for more than a year.

The Chinese fleet of Admiral Zheng He in the fifteenth century had 250 ships, including "treasure ships" said to be 440 feet (135 m) long, 180 feet (55 m) wide, and weigh 500 tons. A large rudder post I visited in Nanjing, China, was 36 feet (11 m). It was built to steer a wooden ship more than two football fields long. Even Zheng He's great wooden ships are dwarfed today by the USS *Gerald Ford*. That aircraft carrier, which entered service in 2020, has 25 decks and stands 250 feet (76 m) high on dry land. The Statue of Liberty is 100 feet (30 m) shorter. The ship is more than three football fields long and nearly a football field wide on its flight deck.

From the spread of ideas and human settlements to early trade routes and globalization, ocean shipping has nourished the development of civilization. The introduction of container shipping in the late 1960s revolutionized maritime transport and has triggered an unprecedented surge in ocean commerce over the last 50 years. In 2018, the world's commercial shipping fleet consisted of 94,171 vessels, accounting for over 80 percent of global trade by volume and more than 70 percent of its value. Today, climate change is leading to the emergence of new trade routes that reduce shipping distances and travel times, such as connecting Europe to China through the Arctic Ocean. Ocean shipping is likely to continue to expand for the next century.

When, 35,000 years ago, the first Neolithic humans began showing up on Pacific islands, they brought with them pigs, dogs, and domestic birds and were accompanied by rats and fleas. They hunted local species of mammals, crocodiles, giant iguanas, and birds, sometimes to extinction. As they planted bananas, breadfruit, sugarcane, yams, ginger, coconut trees, and bamboo, they altered the habitats of wild plants and animals, driving many of those to extinction. Nonetheless, archaeological evidence suggests that after the rough period of early colonization, these Pacific islanders learned to live in harmony

SAIL TRANSPORT NETWORK

Recognizing the damage that fossil-powered cargo ships are causing to both atmosphere and ocean, seamen around the world have joined a movement to bring back sail transport. Beginning with small shipments of high-priced items, they have branched out to mixing diverse cargos with selections of scenic port cities and offering sailing tourists the opportunity to embark upon a new type of ecosystem-respectful commerce.

Aegean Cargo Sailing, for instance, relinked the old trade routes of the ancient Mediterranean in 2019 to carry cheeses, pastries, and liquors from Kea; herbs, cheeses, and spirits from Andros; wines and honey from Skyros; organic dried plums from Skopelos; tuna in olive oil, and beer from Alonissos; cheeses and pastas from Lemnos; cheeses, spirits, wine, and olive oil from Lesvos; herbs, cheeses, wines, and mastics from Chios; herbs, wines, and honey from Ikaria; and olive oil from Samos. In February and March 2020, they departed Athens to collect products from Ikaria and other islands to deliver to the Slow Food movement in Italy and Switzerland. After a short layover in Greece, they plan to resume sailing after summer, and then take health-food products from the islands to food expos in Torino and Paris in October. In 2021, a food cooperative in Austria wants Aegean Cargo Sailing to bring them island produce through the port of Trieste.

Agean Cargo Sailing

beside the sea. They could fish, hunt, and gather, more or less sustainably, and create conditions for abundance and contentment. And then, from that stability, they could dare to venture out in search of new and distant lands.

When St. Brendan sailed from the Irish coast westward in the ninth century, he did not know about tides and currents. Indeed, he had never seen a whale before and called it "a sea demon." Luck and their skill at fasting favored Brendan and his companions, who, with a small sail and several oars, were able to get within sight of Iceland and back to Ireland without being swept up to the North Pole or borne southward to the coast of Portugal or West Africa.

When a landmass is encountered in the Northern Hemisphere, the waters veer to the right. In the Southern Hemisphere, they swerve left. This sets up the clockwise gyre of waters in the upper 100 meters of the coastal Atlantic, with the Gulf Stream flowing north and the Azores Current flowing south.

The predictability of the currents provided early Pacific Ocean navigators four main routes: a southern current out away from all the islands, the South Equatorial Current heading westward and slightly south, and two contrary currents above the equator that separated Hawai'i from the rest of Polynesia.

Pacific Islanders did not have a written language or chronometers, but they could navigate north and south by the stars or east and west by the movements of birds and winds and knowledge of these four currents. Experienced Marshallese or Samoan navigators knew their location by studying the patterns of waves, the smell of the air, the color of the clouds, and the feel of his boat. He or she could read the currents the way hunters might track an animal through a forest. A voyage out of sight of land from Tahiti to Hawai'i might have taken three or four weeks, with winds first blowing east to west, then west to east, and finally east to west again. As difficult as it was to navigate that route, stories and songs of both islands, and recent DNA evidence, tell of frequent passages back and forth for many centuries.

Great navigators and racing captains know, too, how to find and use a slippery sea. When a current of freshwater locates itself on top of shallow, saltier water, the freshwater lens can be pushed by the wind to slide over the saltier layer, even if contrary to the direction of the primary current. A skilled boatman can use this slippery sea to accelerate his boat faster than it would normally go.

One extensive ocean current known as the Atlantic Meridional Overturning Circulation (AMOC) is vital for the weather over much of the Western world. It moves warm water (and pleasant temperatures) from the upper layers of the Atlantic to the north (warming New York City, Boston, and Halifax on the eastern seaboard of North America), then crosses over to Northern Europe (warming Ireland, England, Denmark, and the Netherlands). If the AMOC slows, as it is beginning to, temperatures may drop by a few degrees in those places, depending on location. A substantial weakening would cause a decrease in marine life in the North Atlantic, more storms in Northern Europe, less summer rain in the Sahel and South Asia, change in the number of Atlantic hurricanes, and an increase in regional sea level along the polar coast of North America. All across Europe, summers would be cool and winters much colder.

What causes the AMOC to change speed? Water density swings due to both salinity and temperature at the surface of the ocean produce variations in buoyancy, which cause sinking and rising of water, which is the pump that moves the current. Lately, researchers think that the melting of the ice sheets in the Arctic and Greenland may play a significant role in the AMOC, and so may rainfall, or the lack of it, at the equator. While it is unlikely to collapse completely, the chance of the AMOC temporarily shutting down in the next 100 years is around 15 percent. Still, if our greenhouse gas emissions were to continue as they are today, or increase, those odds rise to 50-50. If the AMOC stalls completely, the effect would be to make tropical regions on land much warmer and countries closer to the poles much colder.

Moderating against this are recent Arctic winters that have seen twice the open seawater due to a shrinking ice cap, with older, saltier ice melting. The water around the North Pole is getting saltier. Because it is icy cold and denser, it sinks, sending more energy into the thermohaline circulation to recharge the AMOC.

No one should underestimate how much energy there is in icy, salty water. Between Iceland and Greenland, salty meltwater leaving the Arctic southward narrows into a "jet" of higher energy flow just before colliding with warmer, fresher Atlantic coastal waters. This is one of the few places in the world where the ocean is getting colder. The world's largest underwater waterfall—175 million cubic feet of water per second, equal to 2,000 Niagara Falls—plunges unseen down 11,500 feet (3,505 m)—three and a half times the drop of the world's highest terrestrial waterfall, Angel Falls in Venezuela—into the Denmark Strait Cataract, swings around the southern tip of Greenland, and curves up into Baffin Bay, Labrador. From there, it is on to the Flemish Cap off Newfoundland, where the nutrient-rich waters of the Labrador Current pass below nutrient-poorer waters moving north on the Gulf Stream. Dense fog arises from that contact, and from this fog emerges one of the most productive fishing grounds in the world, and the setting for the film *The Perfect Storm*.

Oceanographer Rob Moir set out to learn how the climate emergency is changing temperatures below the thermocline. He enlisted the help of some male narwhals, who, as he explained, "carry a five- to ten-foot-long tusk that erupts through the lip on the left side of the upper jaw. What was once a canine tooth has become a most remarkable left-handed helix spiral. Millions of nerve endings in the tusk connect seawater stimuli with the narwhal's brain. By rubbing tusks together, male narwhals are thought to communicate water characteristics each has experienced." Moir set up an experiment:

> Narwhals dive to the ocean floors of the Baffin Bay for Greenland halibut ten to twenty-four times a day. Fourteen narwhals were captured and fitted out with satellite-linked, time-, depth-, and temperature-recorders. When they surfaced, data was transmitted to the researchers. CTD-Rosettes were used from ship and helicopter to verify the accuracy of what the whales were finding.
>
> One of the deepest diving cetaceans in the world, narwhals proved to be excellent "ocean samplers." Dives lasted more than twenty-five

minutes. Given the distance underwater involved, dives were vertical. Deep vertical dives are ideal for repetitive depth and temperature
casts. Narwhals also favor diving for fish beneath Arctic offshore ice.
For obvious reasons, these are areas where few oceanographic studies
have been done. Much of Baffin Bay is covered by ice in various forms
throughout the summer.

Tags put on narwhals in August and September lasted up to seven
months before falling off. Data was transmitted with every surfacing
and every breath of the whales. Data was collected well into the winter months.

Narwhals documented that the surface water, in contact with
overlying sea ice cover, was thinning. The thermocline had moved up,
and the surface water was 160 to 260 feet [48.7 to 79 m] less deep
than previously observed. Beneath this relatively cool water is a much
larger warmer water mass, the West Greenland Intermediate Water.
Narwhals found that the depth of cool surface water is lessening, and
warmer waters below are expanding upwards.

By teaming in this way with deep-diving cetaceans, Moir was
doing more than merely gathering data to confirm what scientists
have already suspected about ocean warming. He was showing that
species can partner with each other, and maybe just help us both get
out of this alive.

Humans, with their big brains and ability to cooperate and fore-
cast their future, have had an outsized effect on nature as they migrated
to the farthest habitable regions of the globe. They caused extinctions
big and small, some intentional—as in the hunting of the gray wolf
or the American bison—but most unintentional. If you consider the
difference between Brendan's boat and the USS *Ford,* it is easy to

imagine how rapidly the growth of human technology has magnified our species' impact on the rest of nature. Our challenge is to turn this force to good—to use our intelligence, and the intelligence of Gaia—to restore ocean ecosystems to at least the point where they can regenerate themselves, and then, hopefully, allow them to support us in the way they have always wanted to, sustainably.

THE THREE MAIN FORCES

Three main forces direct the movement of water in the ocean: waves, currents, and tides.

Waves

The spin of the Earth, the push of winds, the rumbling of earthquakes and volcanoes, and the movement of landslides create waves. Most of the waves you see are the shorter, less intense variety that occurs at the surface, and, most of the time, those are caused by winds blowing across the water. Sometimes other forces create higher, more massive waves like tsunamis, or tidal waves, that can bring considerable damage on land, or rogue waves that happen at sea and can overturn boats. These larger waves, which may be moving unseen below the surface, are most often caused by underwater earthquakes that create shifts in the ocean floor.

As waves approach the shore, the area available to contain their energy shrinks because of the slope of the seafloor, which causes the waves to rise higher, so that they break rather than lap ashore quietly. If the ocean floor rises smoothly and evenly on the waves' approach to land, the wave tips may curl before they break. Evenly breaking curls are what surfers look for to find the best waves.

Currents

The world rotates west to east. Because of this, winds that blow toward the North Pole track slightly to the east. This bending to the right is called *the Coriolis Effect*. Out on the ocean surface, the water feels the Coriolis force and drifts a bit to the right of the wind. The water below that water flows a bit more to the right, increasing rightward pressure with depth, each layer with more spin effect than the layer above. That motion is called the *Eckman Spiral*. At the surface, water responds more to the wind than to the Coriolis force. This wind-directed motion sets up a shear across the spin-directed waters beneath.

A continuous stream of water moving along a definite path is called a current. It does not have to move horizontally along the surface; it could be an upwelling current from below or a descending current from the upper reaches to the depths. Wind, tides, density gradients, and planetary motion drive currents. They can be warm or cool, salty or fresh, and that can make them rise or fall with the surrounding water since heat rises above cold and cold falls below heat, and since freshwater, being less dense than salty water, rises while salty water sinks. For example, water coming from the Arctic is colder than water from the equator, so the cold water drops below the warm water as they pass each other. This creates an enormous motion within the oceans known as a conveyor, or *thermohaline*. The movement is powered by this temperature change, as the water warms from the sun at the equator and rises, then cools at the poles and sinks.

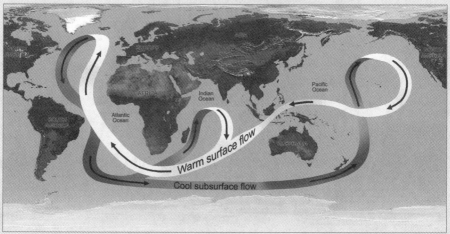

NASA/JPL

Tides

The gravitational pull between Earth and the Moon causes tides—the rising and falling of the ocean in relation to land. The highest tides occur when the Moon is most full, which is when the water rises the most, but tides can change to high or low each day, or even more than once each day, depending on your location.

PART TWO

CHAPTER

3

Emptying the Ocean

As our ancestors began to venture into and upon the ocean, they started to have effects greater than their small numbers might have predicted. Early attempts to catch fish likely involved hand gathering, then spears, nets, and hooks. Stone Age Neanderthals fished and drew paintings of it on cave walls. Bone samples of Tianyuan man, a 40,000-year-old modern human from eastern Asia, provide evidence that he regularly consumed fish. Early humans made fish hooks from bamboo, wood, bones, and shells, then from iron, copper, and bronze. They made lines from vines, horsehair, silk, and catgut.

Slowly at first, but then speeding up, fishing and the technology to support it grew. By the time my ancestors 12 generations removed left England to go to North America in the early seventeenth century, fishing trawlers were towing nets in the deep ocean. The first steam-powered trawlers appeared in the 1870s, when my ancestors were traveling across the North American continent in covered wagons. After World War II, boats used radio navigation and fish-finding sonars. A single trawler could pull 60 tons of fish from the sea every time it lifted its net. Today there are some four million commercial fishing boats, and 40,000 of them are more than 100 tons. Two generations ago, only about 20 percent of the fishing vessels of the world had motors. Today, a fleet of small boats boasts as much engine power as large trawlers. Commercial fishing gear includes weights, nets, seine nets, trawls, dredges, hooks and line, lift nets, gill nets, entangling nets, and traps.

The footrope of a bottom trawl is designed to run close to the seabed. Sometimes it is weighted and just drags, and other times it may use rollers or wheels to move along the bottom. It goes after fish that linger near the bottom, such as plaice, sole, grouper, and flounder, but will also catch noncommercial fish, or bycatch, such as manta rays and moray eels, without specifically targeting them. In many beam trawls, there are "tickler chains" set ahead of the net mouth to scare up fish that hug the bottom. Moving along at the speed of the trawling boat, footropes and chains slice off or bruise sea fans, corals, and sponges and chop down whole meadows of seagrass and forests of kelp in search of their prey. Sometimes they catch large boulders and roll them across the reefs, breaking apart huge coral chunks. Six million square miles of ocean are fished this way every year. Some of the same areas of reefs are fished five or more times every day. Bottom trawlers cause more than 15,000 square miles of dead, damaged, and dying bottom life every day, an area the size of Europe or America every year.

We have no records of fish catches before the last decade of the nineteenth century, but early photographs show cannery boxes overflowing with fish, and fish too large to box—halibut, cod, turbot, and tuna—piled high on the docks. We can see shores on the Puget Sound piled waist-deep in salmon.

In 1870, the shad were so numerous in New Jersey at the end of May that people would gather them with pitchforks and fill horse-drawn wagons when the tide went out. A 1913 report on menhaden,

an oil fish, said that over a billion fish were harvested in Cape Cod that year, yielding 6.5 million gallons of oil and 90,000 tons of fertilizer. In 1915, 400 million pounds of salmon were taken from Alaskan waters. A sailboat in the Gulf of Mexico could expect to gather 2,000 red snappers in a single day.

Menhaden, salmon, and red snapper are all overfished in those same waters today. Each year, fishing commissions set quotas that are well above replacement levels, and the decline deepens. Marine biologist Callum Roberts calls it physician-assisted suicide. In 2010, the status of 275 commercial fish stocks (out of 528 examined) could not be determined, meaning those species were too rare to be counted in the census.

Modern high-tech fishing ships, equipped with powerful engines, refrigeration, sonar, satellite navigation, and mile-long nets, work 17 times harder to catch the same number of fish as a sailboat with rods, lines, and hooks in the 1880s. For every hour spent at sea, these twenty-first-century technology platforms land just 6 percent of the fish hauled in an hour by a nineteenth-century sailboat. Few of us, including the fishermen, see this change. We might recall more fish in earlier decades of our lives, but we have no intergenerational memory to grasp the full picture.

Today, the average person in the world eats more than twice as much seafood every year (about 46 lb, or 21 kg) as when Xen's parents were born, and that rate is increasing twice as fast as is the human population. Today, more than 90 million tons of fish are caught in the wild, and another 90 million are grown in fish farms. It might be comforting to think that, were it not for fish farming, which has grown as rapidly as fish eating has, many more wild fish would be in danger of extinction, but that is incorrect. Some fish farms are stocked with fry

or young, like shrimp or bluefin tuna, that are taken from the wild. Many kinds of desirable food fish are carnivores and must be fed wild fish. Because about 90 percent of the embodied energy of any animal is lost at each step along a food chain, it takes many pounds of wild fish to produce a single pound of farmed fish. Penstock-raised bluefin tuna, for instance, might grow to be 20 pounds (9 kg) but must be fed 400 pounds (181 kg) of wild-caught fish to reach that weight. Rather than sparing wild fish by growing them in farms, we are actually depleting wild fish faster that way.

The sustainability of fisheries is essential to the livelihoods of billions of people. Seafood accounts for 17 percent of all animal protein consumed in the world. The ocean is also an essential source of income—nearly 60 million people work in fisheries and aquaculture, and an estimated 200 to 500 million jobs are directly or indirectly connected with fisheries. For some countries, the fish trade generates more income than most other food commodities combined.

Fifteen years before Xen was born, 90 percent of commercially pursued species of fish were fished at biologically sustainable levels. Today, just 23 years later, 90 percent are either fully exploited, overexploited, or depleted. Each year, the number of overfished species rises, and the number of sustainably fished species goes down. The Mediterranean and Black Seas have the highest rate of unsustainable fishing, closely followed by the southeast Pacific and southwest Atlantic.

In 10 years, Xen may not be able to eat fish as often as people do today. Looking at the rate of decline, if we follow what happened between 1970 and 2020, we can calculate the halving time of ocean fish at approximately 50 years. But remember, 1970 was before computers, navigational satellites, and sophisticated fish-finders. It was before the era of mile-long, practically invisible, nylon monofilament nets towed behind 10,000-horsepower factory ships. If that 50-year decline rate were to be sustained to 2070, when Xen is 58, fish stocks would be one-quarter of their numbers today. That is well below the recovery rate for many species.

Of course, humans are not the only animals to eat fish. When we catch fish, there are fewer fish to feed other fish, sea birds, seals, and polar bears. To fill the demand for seafood, we are now harvesting Arctic krill to press into cakes and fish sticks. Blue whales, the largest surviving marine mammal, weigh around 150 tons and grow to be

100 feet (30 m). They have hearts the size of golf carts. Their primary food is krill, about four tons per day per whale. Humpback whales, best known for their songs and aerial leaps, also migrate long distances to get to the Arctic regions for the krill season.

As we eat the best-tasting fish to exhaustion and work our way to the bottom of the food chain, we starve out the larger fish that need those staple foods. According to Ken Norris, lead author of a study by World Wildlife Federation and the Zoological Society of London that tracked 5,829 populations of 1,234 species, such as seals, turtles, dolphins, and sharks, "Billions of animals have been lost from the world's oceans in my lifetime alone. This is a terrible and dangerous legacy to leave to our grandchildren." The marine life of the ocean is now at the "brink of collapse," he said.

Combining the terms "nutrition" and "pharmaceuticals," *nutraceuticals* are foods containing bioactive molecules with health benefits that extend beyond nutritional value. Marine nutraceuticals, such as omega-3 fatty acids from krill, represent an increasingly large portion of the global nutraceutical market, expected to reach $385 billion during 2020.

Possibly the shortest food chain in the world is this: sunlight and seawater feed phytoplankton. The phytoplankton cells are eaten by other plankton and by krill, which feed hundreds of species of fish, some 10 million tons of squid (which eat 15–20 percent of their body weight every day), more than 40 kinds of birds, and seals, dolphins, penguins, turtles, and the great whales. Fewer krill, and all those species go hungry.

A whale shark fin recently sold for $50,000 in Asia. Whale sharks are an endangered species that are on international lists for protection. And yet, we can find frequent examples around the world of

- fishing without a license or quota for certain species,
- failing to report catches or making false reports,
- keeping undersized fish or fish that are otherwise protected by regulations,
- fishing in closed areas or during closed seasons,
- using prohibited fishing gear, and
- conducting unauthorized transshipments (e.g., transfers of fish) to cargo vessels.

The new UN Sustainable Development Goals call for ending over-fishing and destructive fishing practices by 2020 and restoring stocks "in the shortest time feasible," but in many areas and for many fish stocks, there are no conservation or management measures planned. Even where fishing activities are in managed areas, they may be conducted improperly with impunity by vessels without nationality, or by those flying a flag of a State or fishing entity that is not a party to the regulation or the conservation measures of the host nation. And so it goes. You can call for change, but can you really enforce it?

And after the fish are gone, to where shall we turn? Half of all food-insecure countries are experiencing decreases in land crop production—and so are the affluent industrialized countries in Western Europe, North and South America, and Australia—due to rapid climate change.

We chastise meat eaters for negligent and profligate consumption to the detriment of soils, forests, and climates. However, we say little or nothing to those who consume fish ensnared by dubious practices or from species poised at the edge of extinction. We don't tell the fishermen of the world to stop fishing. In fact, in many places, we train them and pay them to fish more efficiently. And they will.

Careless of the consequences, we are converting coral reefs and seagrass meadows into "endless, monotonous expanses of shifting gravel, sand, and mud," says Callum Roberts. Fleets sent out by Spain, Taiwan, Sri Lanka, and Japan fished out the breeding grounds and migratory routes of tuna, swordfish, mahi-mahi, and sharks. Now they are chasing Chilean sea bass, pollock, and king crab to the polar ice shelves, scouring everything to 10,000 feet (3,048 m) down. The United Nations reports that two-thirds of commercial fish populations have collapsed. Some major fisheries, such as Atlantic bluefin tuna, are reduced by 95 percent. And yet, as technology improves and the human population grows, boats still set to sea to try to find and harvest that last 5 percent.

The UN predicts that "total world fish production (capture plus aquaculture, excluding aquatic plants) is expected to continue to expand . . . to reach 201 million tons in 2030." Against that prediction are not only the dwindling odds of finding large enough fish to harvest but also the changing climate of the ocean. The primary production in the tropics, meaning its ability to make food from sunlight and nutrients, will decline—best estimates are 11 percent—during Xen's lifetime. Eleven percent may not seem like much, but that is *primary* production, upon which all secondary and tertiary feeders depend.

In *Moby-Dick,* the crew on the deck of the whaling ship *Pequod* looked down into the oceans and met the eyes of their prey:

> Far beneath this wondrous world upon the surface, another and still stranger world met our eyes as we gazed over the side. For suspended in those watery vaults, floated the forms of the nursing mothers of the whales, and those that by their enormous girth seemed shortly to become mothers; . . . and as human infants while suckling will calmly and fixedly gaze away from the breast, as if leading two different lives at the same time; and while yet drawing mortal nourishment, be still spiritually feasting upon some unearthly reminiscence; . . . even so did the young of these whales seem looking up towards us, but not at us, as if we were but a bit of Gulfweed in their newborn sight.

Melville acknowledges the sentience and spiritual lives of the whales, whose existence the crew of the *Pequod* is ending in order to light lamps in Nantucket, and compares his prey to nursing mothers, but then he wraps his observation in a caveat that while they seem sentient, they are really just staring mindlessly. They are not human, after all.

This failure to lend thought to the unique intellectual capacities of other species has allowed us to hunt to extinction the Steller's sea cow (27 years after Europeans discovered it), sea mink (for its fur), Labrador duck (for its eggs), New Zealand grayling (by deforestation), Canary Islands oystercatcher, Caribbean monk seal, Saint Helena large and small petrels, Pallas's cormorant, eelgrass limpet, Great Auk (the only species in its genus, killed for feathers to make pillows), and Japanese sea lion (for body parts used in traditional medicine).

Only now, when so many are already gone and more seem destined soon to go extinct, are we beginning to grasp the distinctive forms of sentience and unique contributions that each makes. We have only recently developed sonar, but dolphins, porpoises, orcas, sperm whales, or their ancestors have used it for 30 million years. Octopuses have the highest brain-to-body mass ratios of all invertebrates—greater than that of many vertebrates. Unlike vertebrates, the complex motor skills of octopuses are not organized in their brain but are distributed and semi-autonomous. The octopus's suction cups are equipped with chemoreceptors so the octopus can taste what it touches. I think it is safe to say we have not even begun to understand the intelligence of an octopus, or, for that matter, a blue whale.

One thing we can say, however, is that Lynn Margulis and James Lovelock, in their Gaia hypothesis, were likely right when they theorized that all these intelligences across all genera of life combine to function as a single, self-regulating system—to make a higher intelligence of the whole. If we now carve away pieces, in larger and larger chunks every year, we are lobotomizing our planetary brain.

When human numbers grow so great, and the fish we are catching so small, both in numbers and in size, that we are starting to haul and eat krill, there is something very wrong. It does not have to be like this. We can learn and teach our children the truth about limits. Living within the limits of the ocean's bounty, we could all live very well indeed, and so can the octopus and the whale.

THE ORNAMENTAL CATCH

In 1369, the Hongwu Emperor of China established a company that produced large porcelain tubs for goldfish. Ever since then, marine organisms have been kept as pets or traded as ornamentals. We hang them on our walls, wear them on pendants, and keep them in desktop aquariums.

Today, exotic fish from the rivers and seas of more than 40 countries fly around the world in jet airplanes in order to fill elaborate tanks in shopping malls and restaurants. A 20,000-gallon (76,000-liter) tropical reef aquarium stands behind the reception counters at The Mirage in the desert oasis of Las Vegas. Millions of marine organisms will be removed from the ocean this year and inserted into private and public aquaria worldwide—over 1,800 species of fish and hundreds of species of corals and invertebrates.

Unlike freshwater species that are mostly farmed, nearly all saltwater aquarium fish are wild-caught, primarily from coral reefs. In recent years, aquarium hobbyists have shifted from fish-only tanks to miniature reef ecosystems, increasing demand for live corals and plants. The largest aquariums may have otters, turtles, dolphins, whale sharks, and orcas.

The complexity of the industry, absence of regulations, and lack of traceability prevent reliable estimates of scale, or of destructive collection practices, overharvesting, and depletion of rare species. In 1993, the United States Census Bureau reported that 10.6 percent of US households owned ornamental freshwater or saltwater fish, averaging 8.8 fish per household. Approximately 200 million fish worth $44.7 million were imported into the United States in 1992. These fish comprised 1,539 different species, a total aquarium fish population of approximately 85.7 million. These numbers are likely to be higher today.

Fish are sometimes maintained in inadequate conditions by inexperienced aquarists who often keep too many fish in one tank or add fish too quickly into an immature aquarium, killing many of them. Counting only imported fish, the US aquarium fish population turns over more than 2.3 times per year. Some people give their fish body piercings, while others tattoo them with company or sports team logos.

Genetically modified (GM) fish, such as the GloFish, are likely to become increasingly available, particularly in the United States and Asia. The GloFish is a patented and trademarked brand of fluorescent zebrafish now sold in many fluorescent colors—"Electric Green," "Sunburst Orange," "Moonrise Pink," "Starfire Red," "Cosmic Blue," and "Galactic Purple." A GM GloFish Rainbow Shark recently joined the lineup.

In theory, reef fish should be a good example of a renewable resource that encourages fishers to maintain the integrity and diversity of the natural habitat; more and better fish can be exported from pristine habitats than those that have been polluted or overharvested. However, this has not been the case. Fish are overfished or carelessly caught by many destructive practices. Cyanide is dumped in the water to avoid the problem of netting or trapping among corals. Cyanide can irreversibly damage or kill both the target fish and other fish, mammals, reptiles, invertebrates, and corals. But even without cyanide, mortality rates during collection and shipping are high.

In January 2011, the Maui County Council became one of the first regulatory bodies in the world to pass a measure requiring aquarium fisheries to adopt humane practices. The Maui standards prohibit clipping the fins on fish to protect plastic shipping bags, puncturing swim bladders that fish use to regulate their buoyancy, which enabled divers to surface rapidly; and "starving" the fish before shipping to allow smaller shipping bags. The measure also requires shippers to file mortality reports on the animals they ship.

While the Maui regulations are an excellent example of what can and should be done, they are limited in their jurisdiction and very few other authorities have adopted them. Well-paid commercial interests trump conservation. The best leverage we may have is for purchasers of exotic fish to insist on sustainable and humane catch certification before making any purchase. Or don't buy an aquarium.

4

Blizzards of the Deep

Ever since 1932, when biologist William Beebe and engineer Otis Barton squeezed through the 14-inch (36-cm) opening in their hollow metal ball they called a bathysphere, humans have been exploring the depths of the ocean. Beebe and Barton reached 2,200 feet (670.5 m) on that journey and would descend to 3,000 feet (914.4 m) two years later. They saw flashlight fish, with bean-shaped pouches below their eyes filled with bioluminescent bacteria that blinked on and off as the fish winked. They saw sea jellies that made light by reacting a chemical called luciferin with oxygen. Other fish had eyes shaped like tubes to allow them to take in more light, or, like the giant squid, eyes the size of soccer balls. In 1960, Jacques Piccard and Don Walsh, in their research vessel *Trieste*, set a record that can never be broken by touching down on the lowest point on the surface of the Earth, 35,814 feet (10,916 m) below sea level, in the Mariana Trench.

While the early explorers could take photographs through a porthole or describe what they saw, today's research vessels have robot arms, sensors, and video cameras to gather samples and show what they've seen to the world. We have discovered countless organisms like the vampire squid, named for the webbing between tentacles that resemble a red batwing, and a female octopus that sits on its nest of eggs for four and a half years as its babies get large enough to enter their dark world.

Sunlight filters through the uppermost layers of water, but from 650 feet (198 m), it encounters what oceanographers call a twilight zone. Below that, from about 3,280 feet (1,000 m), is the midnight zone, with no sunlight, temperature close to freezing, and enough pressure to crush a golf ball. For every 33 feet (10 m) you descend, the pressure increases by 14.7 pounds (6.6 kg) per square inch. Animals like sperm whales and sea lions have flexible skeletons that collapse to accommodate the pressure when they dive.

Invertebrates—animals that have no skeletons—reign on the ocean's floor. With bodies or shells that maintain the same pressure inside as outside, there are deep-sea versions of starfish, sea urchins, sea cucumbers, sea anemones, and even corals that do not require warmth or sunlight. Mud-dwelling ratfish have skeletons made of cartilage and can find food like worms and clams by sensing electrical fields. Perhaps this explains why sharks and rays were able to survive so many extinction events: Their special abilities—to contract, sense electrical fields, and smell at a great distance—allow them to dive deep and find food when all of the usual sources have disappeared. Another strategy is that of the giant isopods, which can slow their heart rate and go for years without eating.

Some creatures of the deep rely on whale falls for food: When a whale or other large fish or mammal dies near the surface, its carcass drifts to the bottom, where it becomes food for thousands of bottom dwellers, sometimes for many years. If there are more deaths of marine animals now from climate change, poisons, oil spills, radioactivity, and other human-caused disasters, it stands to reason these are banquet years at the bottom of the sea.

Other animals, like jellyfish, vampire squid, and sea urchins, rely on what we have come to call "marine snow" for their food—tiny bits of dead plants and animals, feces, and plastic that fall like snowflakes from the surface and upper layers of the ocean.

We used to imagine that heat from the surface would not reach the depths of the deeper ocean during Xen's lifetime, or for many lifetimes afterward. Now we know, from diving narwhals and other sources, that it is getting there rapidly.

So is our plastic snow. When a 2015 expedition to the Mariana Trench took samples of crustaceans on the ocean floor to analyze, they discovered that even those had plastic in their diet. There are now microplastics found in one-third of fish caught and examined. Microplastics have been shown to cross the blood-brain barrier and affect behavior. Reports of fish stranding on beaches, their bellies filled with plastics, are becoming more common.

In 1960, there were 15 million tons of plastic in the world. By 2020, we were adding 400 million tons per year. The plastic entering the ocean is doubling every five years, so by 2050, the world will add 1.2 billion tons per year, barring sudden deindustrialization. Today there is one ton of plastic for every three tons of fish. By 2050 or sooner, there will be more plastic than fish.

And yet, each year, millions of tons of plastic are flushed into the ocean. Sea turtles will eat it as they browse the floating islands of sargassum. Seagulls, terns, herons, and penguins will pick the colorful bits off beaches and feed them to their chicks. Dolphins and salmon will eat smaller fish who browsed plastic from coral reefs. Much of it will fall to the deep seafloor as marine snow. Many of these indestructible polymers are known cancer causers, but the full toxicity of all of them—and the new kinds being introduced every year—is still unknown. For seabirds, whales, and fish that fill up their stomachs with indigestible plastic debris, or sea animals that tangle in abandoned plastic nets, six-pack rings, or floating ropes and fabrics, we don't need to know how toxic they are because these victims will die of starvation, strangulation, and defenseless predation.

There is something uniquely stupid about designing a throwaway item to last forever and to be deadly, generation after generation, even when it falls like fresh snow through the water.

Coral reefs provide food, coastal protection, and income for some 275 million people worldwide who live near beaches, but they are under stress from climate change, pollution, overfishing, and the billion plastic items ensnared in them—including diapers, tea bags, shopping bags, fishnets, bottle caps, and toothbrushes. This is as true in remote reefs as in those close to cities. When corals come in contact with toxins and plastics, the risk of contracting disease rises from 4 percent to 89 percent. That is because corals have fragile tissues that can be cut and wounded, and those wounds are easily infected if they are in contact

STRAW WARS

The young woman strode past the three judges and spoke directly to the standing-room-only audience of more than 1,000. "I am Chelsea Briganti, my company is Loliware, and we are part of a plastic-free movement. We believe that single-use plastics should never be designed to last forever. They should be designed to disappear."

This was the long-awaited final pitch-off between what had begun as a large flock of eager entrepreneurs two days earlier. I was attending the Collision Conference at the end of May 2019, when it was held in Toronto for the first time. I had already listened to Chelsea's pitch twice before, each time with three new judges, as she advanced through the competition, and it was clear to me she was getting better each time. The first time she had 15 minutes and was able to use a larger slide deck to tell her story with images and numbers. This final time she had only five minutes and just a few slides, and her competitors were the two very best other finalists from the entire contest. One was a Silicon Valley gourmet vegan food-by-mail service. The other was a wireless credit bank in rural Africa.

Briganti had seemed nervous in some of the earlier competitions, but this time she was fearless. "There are currently five trillion pieces of plastic floating in the ocean. I've seen it firsthand. I was at the Ocean Plastics Leadership Summit down in Bermuda last week. I was diving in the open ocean, and I was surrounded by plastics. It was like being in a snow globe.

with poison or an item covered in all sorts of microorganisms. Plastics made of polypropylene—such as toothbrushes and bottle caps—can become "heavily inhabited" by bacteria that are associated with a group of coral diseases known as white syndromes.

"You could be diving, and you think someone's tapping your shoulder, but it's just a bottle knocking against you, or a plastic trash bag stuck on your tank," reef researcher Joleah Lamb told the Reuters news agency. "It's really sad."

In January 2020, the nation of Palau became the first to ban "reef-toxic" sunscreens to save its corals. One of the side benefits is that those little plastic tubes of sunscreen that people use and discard will also vanish from Palau. Thanks to grassroots campaigns like The Last Plastic Straw, California became the first US state to implement a ban on plastic straws. Dine-in restaurants are no longer allowed to provide customers with straws automatically. Instead, customers

"And straws are no exception (slide changes to read, '360 billion straws used world-wide annually'). We use them and discard them in the billions annually. They end up in our waterways and across our shorelines. I have to ask myself, why, if we are only going to use something once, have we engineered it to last forever? This thinking seemed fundamentally flawed, so I set out to change it (slide changes to read, 'the straw of the future')."

She gives each of the judges an edible straw made from kelp tinted with blue-green spirulina algae.

"We asked ourselves, what would radical change look like? The straw of the future would be designed to last for hours, not centuries. It would be made from carbon-sequestering kelp, and after you were done using it, it would disappear due to natural processes."

The first wave of Loliware straws started shipping in 2019 to fill Kickstarter preorders and to some early adopter companies she had signed, such as Marriott and Pernod Ricard. "I have demand for 18 billion straws through GBG [Global Brands Group, a joint venture with Creative Artists Agency, currently the world's largest brand management company] for coffee shops and fast-food outlets. For every one billion straws that I replace, there is $100 million in revenue.

"And that's just the beginning. With our game-changing technology, we will be creating and commercializing the future of single-use plastics made from kelp. Imagine all cups, all lids, utensils, plates, and films [used in packaging]—designed to disappear.

"Together, let's create plastic-free oceans once again, for generations to come. Thank you!"

who need plastic straws have to request them. Restaurants that violate the ban will receive warnings first, and repeat offenders will be fined up to $300. Other states and countries have enacted or are considering enacting similar and even more extensive bans. None of these laws would have happened if the people in those places had not first demanded them.

In 2017, the UN Environment Assembly adopted a global goal to stop the discharge of plastic to the sea. In 2018, at the Commonwealth Clean Oceans Alliance meeting in Vanuatu, 52 countries pledged to ban microbeads in rinse-off cosmetics and personal care products and to cut plastic bag use by 2021. Indian prime minister Narendra Modi has announced his intent to eliminate all single-use plastic in the country by 2022. With a population of 1.3 billion and a fast-growing economy, India struggles to manage its vast waste stream and is a significant contributor to global ocean plastic.

In 2018, Disney, Starbucks, Red Lobster, and Lindblad Expeditions all announced they were getting rid of single-use plastics. McDonald's is also planning to phase out plastic straws at their UK and Ireland locations, coinciding with EU proposals to cut single-use plastic. Family-owned Bacardi Limited, the world's largest spirits producer with more than 200 brands and labels, intends to cut its usage over the next two years by a billion straws. Alaska Airlines, American Airlines, United Airlines, the Chicago White Sox, and Dignity Health hospitals no longer offer plastic straws. Danish brewer Carlsberg became the first beer producer to ditch plastic multipack rings that hold beer and other cans together for nontoxic holders made of recyclable glue. The European Parliament voted 571 to 53 to slash single-use plastic across the continent, beginning in 2021.

Loliware's kelp straws add no flavor to either hot or cold drinks, outlast paper straws that can wither only a third of the way through the drink (necessitating three straws each time), and can be reused if washed and dried. Otherwise, they degrade in about 18 hours if left outside. Alternatively, you can just eat it, because—although flavorless—it is both digestible and nutritious. CEO Chelsea Briganti says Loliware straws will cross below the price point of its paper- and plastic-straw competitors by the end of 2020.

It is none too soon. Microplastics in the oceans will soon outweigh all ocean fish combined. Not really a fair comparison because it is being absorbed into all those fish, as well as seabirds, lobsters, octopuses, and all other marine life, so you really should subtract the nonplastic flesh and blood and compare that.

Maybe the best thing about the Loliware straws—the feature that puts them ahead of all the other bio-based alternatives—is that they are rebuilding kelp forests, that, as we shall see, hold the key to solving many of the ocean's problems.

MARINE PERMACULTURE

More than 800 species of marine animals have been documented to inhabit kelp forests. While some of these creatures are microscopic, the occasional gray whale—50 feet (15 m) long and weighing 100,000 pounds (45,000 kg)—can be found basking in the cooler water, catching shrimp-like amphipods and small schooling pelagics darting through the hanging gardens of kelp. These forests are nurseries for mackerel, sea bass, and barracuda. As the giant whales swim slowly over the muddy bottom, they send scurrying brilliantly colored damselfish, well-camouflaged rays, moray eels, and 60 different kinds of rockfishes.

Kelp forests are not just good habitats; they also grow in cool waters and reduce storm wave impact. They can be sustainably harvested for food and biomass, and the more of these forests we make, the more fish in the ocean. Restoring the ocean's kelp forests and seeding new ones defends the shoreline and deacidifies the water.

While the ocean has served a vital role in climate by fixing CO_2 (2.5 kg $C/m^2/y$), there's a limit to how fast marine flora can absorb more. Many scientists warn we are very close to that limit now. While we've been busy studying how to hold land to two degrees of warming, kelp forests and coral reefs live within one degree of mortality.

Kelp forests can reverse almost all of these trends, but at the moment, wild kelp forests are in danger. A 2016 research study, coauthored by 37 scientists, concluded that "kelp forests are increasingly threatened by a variety of human impacts, including climate change, overfishing, and direct harvest."

In eastern Tasmania, sea surface temperatures have increased at four times the average global rate, laying waste to giant kelp, which does best in ranges of roughly 50°F–60°F. As the ocean warmed, the long-spine sea urchin, which generally cannot tolerate temperatures lower than 53°F, moved in. Urchins are controlled by natural predators like lobsters and sea otters. Lobsters had been heavily fished, and otters, which in Tasmania are called rakali, were hunted nearly to extinction for their soft fur and as "water rat" pests because of the damage they caused to irrigation banks and fishing nets. With favorable temperatures and no one hunting them, urchins boomed and ate their way through the Tasmanian kelp. The same story repeated from southwestern Japan to the Aleutian Islands and from British Columbia down to California. In California, sea stars ate sea urchins, protecting the kelp forests. Then, in 2013, a mysterious disease devastated sea stars, and together with unusually warm currents, the urchin population exploded and ate its way through the kelp forests. Populations of sea life that depended on the kelp (abalone, rockfish, and shellfish) and populations that depended upon those (various fishes, bald eagles, and harbor seals) were also decimated.

For the past decade, Brian von Herzen, PhD, has been directing an effort to regenerate life in the ocean by combining two sciences: geophysical fluid dynamics and marine biology. Von Herzen and his science and tech coworkers at the Climate Foundation have developed an approach they call "marine permaculture." Geophysical fluid dynamics uses waves, sun and wind to bring water and nutrients up from the colder, lower levels into the top "mixed layer," where marine biology feeds kelp and seaweed planted on strings of a lightweight latticed structure. "Our vision for the future is self-guided marine permacultures," von Herzen says.

After trials off Woods Hole, Massachusetts, the first small test beds are growing in the Philippines, with more planned for Australia. Soon the entire architecture may be replicated in Zanzibar, where it will help bring back red *Kappaphycus* seaweed, used in recipes, lotions, toothpaste, and medicines. Red seaweed has for generations been farmed and harvested by women on Zanzibar who, without that, had not had their own source of income. Then, in the 1980s, the waters of the Indian Ocean began to warm. Shallow-water temperatures reached 98°F, which is 10 degrees warmer than red seaweed can tolerate. The warmth-induced syndromes like "ice-ice," similar to thermally induced photobleaching of corals. At first, the seaweeds turn white at the tips. Eventually, they die.

Von Herzen's team believes that large kelp arrays planted in deep water at 30 meter depths will cool the surface mixed layer closer to land, favoring regrowth of red seaweed. Kelp farming like this is permaculture because it provides multiple functions within a "cultivated ecosystem." Kelp maricultures serve as primary producers, oxygen sources, ecosystem support, nutrient sinks, and carbon sinks. They can also dampen wave energy, protecting shorelines from erosion—another big problem in Zanzibar. They could temper the effects of climate change, pollution, and biodiversity loss, locally and globally. As Karen Coates, writing for *The Christian Science Monitor,* described it:

> Back on the beach, Makame and other farmers stand waist-high in gentle waves as the tide begins to roll back in. An overcast sky shields the equatorial sun. The women pluck bunches of vibrant red and green seaweed and offer a taste. "It is like you are eating cucumber," she says. Crunchy, salty, and nutrient-rich.

Pyrolysis of marine biomass with carbon capture and storage (PyCCS) is renewable energy from the ocean with the theoretical potential to meet all global electricity requirements, while also drawing carbon out of the atmosphere. The infrastructure is not significantly greater than solar, tidal, or wind energy, but for every kilowatt-hour of power produced, three tons of CO_2 is removed. The carbon withdrawal occurs when phytoplankton, seagrasses, kelp forests, and marine algae draw carbon from the air and water to build plant cell walls. When the green plants mature and/or die, they are harvested and baked to make energy (from the gases that are burned off) and long-lived biochar (from the carbon skeletons of the plants left behind). Sequestering the biochar for soil remediation of farmland, in seawalls, or at the seafloor to assist in building new coral reefs keeps the carbon from returning to the air.

Over the long term, kelp forests can be planted and managed in the sea much in the same way forests are grown on land. Trophic cascades from these forests include edible seaweeds for people, renewed fish populations, cooler coastal waters, reduction of storm surges during hurricanes, re-oxygenation of the ocean, and biomass energy.

The potential climatic benefits of marine permaculture are so great, the authors of a 2017 article in the journal *Frontiers in Marine Science* argue for economic compensation to seaweed farmers in recognition of the role they play in combating climate change. Incentives like that would rapidly expand the industry and women's economic autonomy. Zanzibar represents just a fraction of the global seaweed trade: roughly 30 million tons produced annually, according to the UN Food and Agriculture Organization. The idea is one that could be widely repeated. The team at the Climate Foundation has outlined how the kelp can also be refined into carbon-negative biocrude that could replace fossil fuels for transportation and power generation.

The Climate Foundation says that, ultimately, how much worse it gets for life in the ocean, or how much better, will depend on the choices we all, as individuals and families, make in the next decade. We can continue to destroy life, or we can choose to regenerate it.

PART THREE

CHAPTER

5

Radioactivity

At the end of the nineteenth century, scientists started experimenting with elements that had unstable energies. In 1896, Henri Becquerel discovered that uranium emitted rays that resembled X-rays. Marie Curie suspected that the radiation came from the atom itself. Her work with uranium disproved the conventional wisdom of her time—going back to Ancient Greece—that atoms were indivisible.

Madame Curie was a physicist, not a medical doctor, so she did not recognize the health effects of handling uranium, thorium, radium, and the other radionuclides. Indeed, she suspected the effects would be beneficial. She carried test tubes containing radioactive isotopes in her pockets and stored them in her desk drawer. Although her many decades of exposure to radiation caused chronic illnesses (including near-blindness due to cataracts) and ultimately her death, she never acknowledged the inherent health risks. She likely did not recognize the symptoms when she began to feel weak and lose her hair. After her death, and to this day, her papers and effects are too radioactive to be handled and her laboratory is unsafe to enter.

When extra subatomic particles fly out from a radioactive atom, they are like tiny bullets or missiles—they break genetic codes in cells.

Sometimes that simply kills the cell, as it will most often with higher doses, but at low doses, slight genetic displacements can re-form into mutations, such as cells that are cancerous or birth deformities. Sometimes the re-formed codes are passed along to future generations and can produce hundreds of new and different deformities and diseases. In the 1930s, scientists learned that only about 1 percent of the total effects of radiation appear in each separate generation. The other 99 percent echo for a hundred lifetimes. In St. George, Utah, today, public health clinics get about 140 new patients per year from the genetic legacy of the desert atom bomb tests of the 1950s.

On March 11, 2011, an earthquake in the ocean near northern Japan generated a 14-meter-high tsunami that swept over the seawall at the Fukushima Daiichi Nuclear Power Plant and flooded four operating reactor buildings with seawater, knocking out the reactors and their emergency generators. The reactors shut down and without generators the plant operators could not cool their radioactive fuel. Within hours, three of the reactors melted and exploded, sending parts of their radioactive fuel into the sky, land, and ocean.

When you receive radiation treatment like a CT scan, it is sudden and one-off. The technician presses the button, and it is on and then off. There is no danger from the machine when it is off. When radioactive elements like cesium-137 (just one of the hundreds of elements in a nuclear reactor) disperse into the environment, there is no off switch. Inhaling or ingesting it can kill a person, a dolphin, or a seagull, but then as the individual's body decomposes after death—as bacteria, worms, and fungi eat away the flesh and bone—the isotope goes back into the food chain to strike another individual, and another, and so on. The cesium released during the Fukushima accident was theoretically capable of 112.5 trillion cancers every second, over and over. The danger is limited only by the isotope's half-life— the time it takes for half of it to decay to a harmless element, which for cesium-137 is 30.17 years. Scientists generally use 10 or 20 half-lives to bracket safety concerns, so for cesium-137, "safe" levels arrive in 302 to 604 years (around the year 2322 to the year 2624). This is admittedly an imperfect measurement since any residue, no matter how microscopic, may still be lethal. Cesium is one of 256 radionuclides released during Fukushima, so we would need to calculate quantities, biological effectiveness, and the decay time of each of those to get the full health picture. Other isotopes in the Fukushima fuel include uranium-235, with a half-life of 704 million years, and uranium-238, with a half-life of 4.47 billion years, or longer than the age of the Earth.

At Fukushima, the end of the accident was not the end of the story. In 2013, 30 billion becquerels of cesium-137 were still flowing into the ocean every day from the damaged and leaking reactor cores. Of the four primary elements in seawater—hydrogen, oxygen, sodium, and chlorine—only sodium takes on intense, short-term radioactivity with the addition of a single neutron to its nucleus: Common sodium-23 becomes radioactive sodium-24, with a 15-hour half-life. Other isotopes produced include tritium, or hydrogen-3 (half-life 12 years) from hydrogen-2; oxygen-17 (stable) from oxygen-16; and chlorine-36 (half-life 301 thousand years) from chlorine-35, and some trace elements. To stop this assault on ocean life and our lives, over a period of five years the owner of the plant constructed more than 1,000 tanks to hold contaminated water away from the ocean. In September 2019, the Japanese government announced that more than one million tons were in storage, but that space would run out by the summer of 2022. The government is planning to release those billions of becquerels to the ocean, but not until after the Tokyo Olympics.

In 1946, after the end of the Second World War, the United States was left with a few atom bombs. It decided to use three of them to investigate the effect of nuclear weapons on warships. At Bikini Atoll in the independent Marshall Islands, residents were evacuated and a target fleet was assembled. The 167 Bikini islanders were told their evacuation was "for the good of mankind and to end all world wars." To make room for the tests, 100 tons of dynamite were used to remove corals from the lagoon.

Admiral William H. P. Blandy placed 95 ships inside the lagoon and let a B-29 drop an atom bomb on a target fleet of four battleships, two aircraft carriers, two cruisers, thirteen destroyers, eight submarines, numerous auxiliary and amphibious vessels, and three surrendered German and Japanese ships. Onboard were 60 guinea pigs, 204 goats, 200 pigs, 5,000 rats, 200 mice, and grains containing insects to be studied for genetic effects. The test was witnessed by 42,000 men and 37 women.

Shot Able sank five ships. The airburst was purposely detonated high enough in the air to prevent surface materials from being drawn into the fireball. With 90 ships surviving, Admiral Blandy was pleased with the result.

At Shot Baker, the weapon was suspended from a landing craft and detonated in 90 feet (27 m) of water, halfway to the bottom of the coral lagoon. When the detonation began, a rapidly expanding hot gas bubble pushed against the water, generating a supersonic hydraulic shock wave that crushed the hulls of nearby ships as it spread out. Eventually, it slowed to the speed of sound in water, which is one mile per second (1,600 m/s), five times faster than sound travels through the air. From the observer ships, the shock wave was visible as the leading edge of a rapidly expanding ring of dark water, called the "slick." Close behind the slick was a visually more dramatic whitening of the water's surface, called the "crack."

Four milliseconds after detonation, the gas bubble broke the surface and started a supersonic atmospheric shock wave. The bomb lifted two million tons of spray and seabed sand into the air in the first second. After it reached 6,000 feet (1,829 m), the top of the cloud became a "cauliflower," and all the water went back down, into the lagoon. At 12 seconds after detonation, a tsunami wave 94 feet (29 m) high passed under the fleet. One second later, the falling water created a 900-foot (274 m) "base surge" that rolled over the ships. Of all the bomb's effects, the base surge had the most significant consequence because even the ships that survived were painted with radioactivity that could not be removed.

Within the first several days, after intensive deck scrubbing by thousands of bare-chested seamen, tests showed that decontamination was not working. More than half the 320 Geiger counters shorted out as soon as they were turned on. Stafford Warren, MD, the chief health physicist, warned the admiral that the Navy would become liable for lawsuits unless the men were evacuated. Warren showed Blandy an autoradiograph of a fish—an X-ray picture made by radiation coming from a lagoon-caught fish. Blandy halted the exercise and classified the reports, and Congress made it illegal for so-called atomic veterans to hire lawyers or sue the government.

That was not the end of Operation Crossroads. At Bikini, 42,037 people had been exposed to plutonium contamination with no protective clothing. Over the next decades, hundreds would suffer cancers and other diseases and die. If they parented children, their children would carry that burden too. These human guinea pigs may have fared better than the test animals on the ships, but only for a brief period.

The 167 Bikini residents who were relocated to the Rongerik Atoll before the tests were unable to find sufficient food to feed themselves in their new home. The Navy left canned SPAM and water for a few weeks and then failed to return. By January 1947, the Bikinians realized they were facing starvation. By July, they were emaciated. Finally, in March 1948, they were evacuated and sent to the Navy hospital on Kwajalein. From there, they were settled onto another uninhabited island with one-sixth the land area of Bikini with no lagoon and

no protected harbor. Unable to practice traditional fishing and farming, they became dependent on irregular food shipments.

Between 1946 and 1958, the United States detonated another 67 nuclear bombs on, in, and above the Marshall Islands—vaporizing whole islands, carving craters into its shallow lagoons, and exiling hundreds of people from their homes.

At Enewetak Atoll, the United States not only conducted its most massive weapons tests but also experimented in 1968 with biological warfare agents with effects that lingered in the environment for many years. The Navy also dumped 130 tons of soil taken from the irradiated Nevada bomb-testing site there and then exploded an atomic bomb into it.

A few years later, the military sent 4,000 US service members to Enewetak—fresh guinea pigs—to build a landfill for the most radioactive and toxic waste, irradiated military and construction equipment, contaminated corals, and plutonium-laced chunks of metal.

"The Tomb" was built into an unlined coral crater created by a hydrogen bomb. The unprotected and unwarned servicemen took three years to scrape up and haul 33 Olympic-sized swimming pools' worth of irradiated junk from around the atoll, dump it into the crater, and cap the crater with a concrete dome. Six men died during the cleanup; hundreds of others developed radiation-induced cancers and maladies that the US government to this day not only refuses to compensate but also to even acknowledge.

Bob Retmier, who was in Enewetak with Company C, 84th Engineer Battalion out of Schofield Barracks, Hawai'i, said he didn't know he had been working in a radioactive landscape until he read about it in the *Los Angeles Times* in 2019.

"They had us mixing that soil into cement," he told the *Times*. "There were no masks, or respirators, or bug suits, for that matter. My uniform was a pair of combat boots, shorts, and a hat. That was it. No shirt. No glasses. It was too hot and humid to wear anything else."

Retmeir was lucky. He got to go home. The Marshallese don't have that anymore. The Tomb is leaking. As the ocean rises, it is washing more radioactivity into the ocean with every tide. If the western Pacific rises as predicted, its dome will go underwater this century.

On an August day in 2018, tens of thousands of dead fish washed up on the ocean side of Bikini Atoll. Dick Dieke Jr., one of seven temporary caretakers working for a Department of Energy contractor there, recalls the water being uncomfortable.

"It didn't feel good to put my feet in it," he said. "It was too hot." Tests showed 92-degree temperatures 30 feet (9 m) below the surface of the lagoon.

"I've never seen or heard of a fish kill in Bikini," Jack Niedenthal, the Marshall Islands' Secretary of Health and Human Services, said a week after the event. "That's surprising and deeply upsetting."

An independent arbiter agreed to by the United States awarded more than $2 billion in damages to the Marshallese for their suffering and loss. To date, the US has paid only $4 million. No enforcement mechanism exists.

What happens to ocean creatures who ingest radionuclides from leaking nuclear power plants or bomb wastes is not very different from what happened to Madame Curie or the atomic veterans. As the isotopes decay within the body of a dolphin or a coral polyp, they send microscopic bullets hurling through DNA chains, causing tumors, sicknesses, defective offspring, and death for a hundred generations. The chances that a single mutation will produce a beneficial result are less than one in a billion. Radioactivity is, for practical purposes, forever, as we can see just by looking up at our sun, a benevolent nuclear reactor providing us energy from the safe distance of 93 million miles.

6

The Sounds of Silence

In 2019, we learned for the first time how it is that once very rare beach strandings of beaked whales have become common and may eventually lead to the whales' extinction. Scientists taking blood samples from some of the stranded whales found high levels of nitrogen. Nitrogen can cause hemorrhaging and damage to vital organs, and in the whales' blood, it was a sign of "the bends"—what divers call the decompression sickness experienced if they ascend too rapidly from a dive.

How can an animal that lives in the ocean and is used to deepwater dives lasting hours at a time get the bends? Marine scientists linked it to noise pollution.

We tend to think of the undersea world as a quiet place, but apart from a few distant locations, it seldom is. Depending upon where you are, and the time of year, the ocean can be as noisy as any city street or rainforest jungle. Shrimp, fish, and marine mammals all use sound, sometimes beyond the audible range of human ears. They survey their surroundings by echolocation, keeping in nearly constant communication with each other, navigating by listening to sounds of familiar bays and estuaries and crackly ice, and detecting both nearby predators and unwary prey.

The call of a whale, or any loud sound in the water, reflects against the smooth surface of the denser, darker, saltier thermocline with very little loss of energy to absorption. In this liminal zone, the water mass becomes a long-distance sound channel because sound waves can interact with neither the seafloor nor the ocean surface. It is like finding a WiFi hotspot—suddenly, bandwidth becomes much faster.

Since low-frequency sounds travel farther than higher frequencies, a fin whale singing at its 20-hertz baritone has a greater range of communication than a minke soprano with an 80-hertz pitch.

If you descend beneath the polar ice shelves, you may well eavesdrop on the conversations and musical recitals of bowhead, humpback, and beluga whales; walruses; and bearded seals. Just about as far away from civilization as you can get, in the dead of winter, bowhead whales are singing below the North Pole. They are also listening to the sounds of ice, and, because underwater sound travels very far, they can plot their routes to places where the ice is thin enough for them to surface, breathe, and dive again. Like narwhals, they have no dorsal fins because those might make it harder for them to surface and breathe when and where they need to.

These days, if you listen, you can hear fin whales and orcas farther north and out of their typical season. Sub-Arctic species are migrating into the Arctic, increasing the competition for food, possibly introducing diseases or parasites, and adding new voices to the soundscape.

When the North Pole is frozen solid, the underwater Arctic is one of the quietest places in all the world's oceans. When it warms in the spring and begins to break apart, it is a boisterous place to be, although not nearly as noisy as a Navy submarine, heavy arctic trawler, or giant cruise liner.

There are also the deafening, even lethal, "whoomps" of seismic waves as oil- and gas-exploration rigs sound the floor with controlled explosions to make 3D maps of Earth's strata below.

These new sounds shrink the acoustic space in which marine mammals can communicate. The noise increases levels of stress hormones and can disrupt feeding and breeding behavior. It changes swimming and vocal behavior. These are not sounds with which marine animals evolved, but every year they are more difficult to escape. They are loud, and they are alien.

Studying the bowhead beachings, the researchers who discovered the nitrogen in their blood came up with a theory that they published in *Proceedings of the Royal Society*. Long-lived species like the bowhead whale—which the Inuit say can live two human lives—were driven almost crazy by long-range, mid-frequency Navy sonar. They had to do almost anything to escape the pain, so they would dive deep into the ocean, and when they could hold their breaths no more, they would surface and beach themselves to flee from the high-decibel noises that were torturing them.

There are many things we can do to prevent this kind of tragedy. Sadly, the most obvious action—prohibiting those kinds of sonars along whale migration routes—although now mandated by international treaties and court orders, has been ignored by militaries in the name of "national security."

Other steps include slowing down ships that traverse the Arctic because a slower ship is a quieter ship. We can also keep away trawlers and cruise ships in seasons and regions that are important for mating, feeding, or migrating. Banning exploration for oil and gas is long overdue because even if new deposits are discovered, bringing them up would be a climate disaster.

In 1921, a senior radio officer in London was asked to think of a word that would indicate distress and would easily be understood in an emergency. Since much of the traffic at the time was between London and Paris, he proposed the expression "mayday" from the French *m'aider* ("help me"), a shortened form of *venez m'aider* ("come and help me"). "Mayday, Mayday, Mayday" replaced the earlier Morse code SOS as the universal call for help in maritime and aviation voice communications. Today, ship captains and pilots simply declare an emergency.

"Seelonce Mayday" (an English bastardization of the French pronunciation of "silence") demands that the channel only be used by the vessel in distress and emergency responders.

It is now time to call Seelonce Mayday in the deep ocean.

CRUISE CONTROL

In 1996, the late essayist David Foster Wallace described his excursion on a cruise liner as a "special mix of servility and condescension." He exhaustively journaled every event, person, and feeling during a seven-night, all-inclusive voyage. In *A Supposedly Fun Thing I'll Never Do Again*, Wallace said he found the repetition of activities onboard and off so banal as to be infuriating.

The volume of wastes these floating cities produce is large: sewage; wastewater; hazardous wastes; solid waste; oily bilge water; ballast water; and sooty, sulfurous air pollution. Cruise ships can emit as much particulate matter as a million cars every day, and the air quality on deck can be as bad as in the world's most polluted cities. Instead of paying for more expensive but less sulfuric fuel, such as liquefied natural gas, ships are installing "emission cheat" systems that let them meet newer air emissions requirements by capturing and discharging the pollutants from the cheap fuel into the ocean.

A 2018 report in the journal *Nature* said 400,000 premature deaths per year are caused by emissions from dirty shipping fuel, which also account for 14 million childhood asthma cases each year. The emission cheat systems, expected to be installed on as many as 4,500 ships to meet a 2020 deadline imposed by the International Maritime Organization, spare some of those human deaths and asthma cases but transfer the pollution—toxic particulates, sulfur, lead, nickel, and zinc—to the marine environment. Still, the air on deck can be deadly. Researchers found that the air on the upper deck of the *Oceana Rivera*, downwind from the boat's funnels, had 84,000 ultra-fine particulates per cubic centimeter (ppcc), about one-third the concentration measured directly above the stacks. Air quality in London's busy Piccadilly Circus intersection, using the same recording devices, was measured at 38,400 ppcc. That 84,000 ppcc reading is closer to what you might find on a hot day near the center of smog-choked Delhi or Shanghai.

By registering their companies in foreign countries, cruise lines can dodge not only corporate income and property taxes but also labor, environmental, and insurance laws. For the 30-million-passenger, $100 billion industry, the cost of an eight-figure ship can be recouped in as few as five years, after which it's all profits, since the more substantial costs are to the ocean, not the companies. Each ship has the pollution footprint of a small city, nearly unregulated and unpoliced. Cruise ships even have a special exemption from the Paris Agreement that might otherwise regulate their climate pollution.

Pampered passengers produce up to 7.7 pounds (3.4 kg) of consumer waste per person per day, from super-soft toilet paper to plastic water bottles. Each passenger's carbon footprint while cruising is roughly three times what it would be for vacationers on land. Many ships shred their plastic to save space, but some take advantage of the difficulty in monitoring ocean microplastics to discard it with treated sewage and greywater. Because cruise ships tend to concentrate their activities in specific coastal areas and visit the same ports, their cumulative impact on a locality can be significant.

In 2015, the *MV Zenith*, owned by a Spanish subsidiary of Royal Caribbean, dropped anchor near a reef off Grand Cayman. Actually, it was more like *dragged* anchor. Patrick Reilly, writing for *The Christian Science Monitor*, said one scuba diver's video shows "the anchor chain draped across the entire reef, constantly moving back and forth across the reef and causing a lot of damage as it did that." In 2017, *MV Noble Caledonia* ran aground on an Indonesian reef, removing 1,600 square meters—about 17,200 square feet—of coral. Tourism organization Stay Raja Ampat wrote that "Anchor damage from ships like these is bad enough, but actually grounding a ship on a reef takes it to a whole new level." Were it not that corals are declining worldwide due to climate change, the reef might repair itself in 100 years. Now the damage is permanent on any timescales humans can fathom. And these incidents are from among only those we know about.

Because many people are nervous tourists in strange countries, cruise lines offer more secure vacation travel. If you don't want the hassle of booking hotels, rental cars, tour guides, and restaurants in unfamiliar destinations with a language you don't speak and a culture you little understand, then for about the same price as an economy flight and all-inclusive resort, you can have the security of familiar, unchanging culture surrounding you.

Tell that to your grandchildren when they ask you about your vacation.

Sea Sickness

Although there were scattered reports in the nineteenth century of diseases in some commercial fisheries, such as sponges and oysters, they mostly went unobserved. As marine biologist Callum Roberts puts it, "If birds fall from the sky, dying dogs litter the streets, or stands of forest trees wither, they are obvious and worthy of attention. But animals and plants in the sea can sicken and die by the millions unnoticed."

A Greenland shark will live 275 to 400 years and extend up to seven meters. Every member of its species has a shrimp-like parasite embedded in its eye. Although this is thought to impair the vision of the shark severely, there is a theory that these copepods are bioluminescent and attract prey. Nicholas Dunn, writing for the Grantham Research Institute on Climate and the Environment, mused:

> A Greenland shark born four centuries ago would have swum in a world that existed not long after the Elizabethan era when the human population was little over half a billion. Now, with the human population nearing 8 billion and the climate changing faster than ever before, I wonder what a Greenland shark born today will experience if she survives to 400?

If she survives even another 40 years (hundreds of tons of Greenland sharks are still caught accidentally by shrimp and halibut trawlers around the Arctic), she will find the company has changed. There will be, for instance, no more Yangtze giant softshell turtles because the last known female died in a Chinese zoo on April 12, 2019. There will be no more vaquita porpoises in the Gulf of California (only 30 remained in 2017; 19 in 2020); no more Hawaiian monk seals, Guadalupe fur seals, Steller sea lions, or southern sea otters. Stocks of wild tuna, mackerel, and bonito have fallen by almost 75 percent since 1970. In 2019, an ingested plastic bag starved a young, three-meter True's beaked whale off Florida's east coast. The population of beaked whales is too small to count.

Each of us now consumes, on average, 10 grams of microplastic per week, about the weight of a credit card. By 2040, each human baby born will have twice the detectable microplastics in its blood as now, and in 20 years, its child will have twice that much, then double that, then twice that again as we complete this century. That is the arithmetic that caught up to that young beaked whale.

In 1979, disease suddenly wiped out nearly all the staghorn and elkhorn corals in the Caribbean. One variety of diseases affecting elkhorn coral, white spot, or *acroporid serratiosis*, was later linked to a sewage bacterium, *Serratium marcescens*, but the "white band disease" pathogen that killed off staghorn and elkhorn corals in 1979 has never been identified and is not the same pathogen. The Caribbean is now being devastated by new versions of this disease.

The outbreak was one of many that have collectively destroyed 80 percent of living corals in the Caribbean since 1977. A herpes virus that exterminated all Australian pilchards was transmitted from host to host through the Southern Ocean at 6,000 miles per year. It likely first arrived in Australia in frozen fish bycatch being fed to aquaculture farms. A bacterial infection passing from the Pacific to the Atlantic through the Panama Canal, likely in ships' ballasts, nearly destroyed all the long-spined sea urchins in the western Caribbean. The pathogen finally died when it ran out of hosts.

Green and leatherback turtles are being driven to the brink of extinction by a combination of ingested plastics, toxic pollution, loss of breeding beaches, and papillomavirus that has been linked to algae blooms fed by urban pollution. In 1988, half the harbor and grey seals in the North Sea died from a distemper virus similar to that found in dogs. Eighty to one hundred thousand seals were struck by dog distemper in Lake Baikal. It took 22 years for the seal population to recover, but then it was struck a second time by the same virus, killing 20,000 North Sea seals. A parasite from domestic cat feces carried in river runoff attacked California sea otters, causing brain inflammation, which didn't usually kill them but weakened them so that more were caught by sharks. The same cat toxoplasmosis affecting the otters has now been seen to have infected clams and mussels, spinner dolphins, and beluga whales.

If you were to visit some Pacific islands, you might be surprised to see a cook place a silver coin on the guts of a fish before placing the filet in the skillet or on the grill. If the silver turns color, they throw the fish away! The cook is testing for ciguatoxin, which is not a parasite but a poison that moves from herbivore fish to those that eat them. It gets its name from the cigua snail, which infected a crew member of a seventeenth-century slave ship. In 2020, 10,000 to 50,000 people will get the poisoning.

For such diseases to become so lethal and widespread, you need a source of poison or pathogen, a vulnerable population of victims, and a way to transfer the poison or pathogen from victim to victim. We know from our own experience that people who are stressed become more susceptible to disease because their immune systems become compromised. We see that bruised and broken corals do not heal quickly and that injured tissues of fish are much more readily infected than healthy ones. In much the same way, cumulative stresses compromise all life in the oceans.

In normal times, sea parasites are kept in check by predatory viruses, limits on their ability to leave particular temperature or nutrient zones, and factors that evolved to create a harmony or balance over millions of years. When things change suddenly, so can these limiting factors. Warmer water may allow parasites to find new feeding grounds closer to the poles or farther into deep water. The sudden absence of a limiting condition may allow parasites to expand their populations. New weaknesses of their prey will also benefit them, allowing them to colonize animals that might have had relative immunity before.

Sometimes the cure can be worse than the disease. To combat ciguatoxin, medical researchers needed to isolate samples of it to test cures. It takes 847 kilos of moray eel liver to separate just 0.35 milligrams of ciguatoxin—not the whole eel, just the eel's liver. To extract livers from so many moray eels is decimating the eel populations of the world's coral reefs.

Of 108 reef systems, only 16 remain untouched. Marine biologist Andrew Caine says, "Together with tourism, biomedical extraction, mining, and the aquarium trade, to mention just a few, coral reefs will soon be on par with the rate of rainforest destruction." Chances are, they already are.

The cure also seems worse than the disease when we acknowledge all the human pharmaceutical waste that has been entering the ocean in recent years. Consider the case of the tiny shrimp known only by its Latin name, *Ampithoe valida*. We can call her Amy. What we know about Amy is that she is native to the east coast of North America from Florida to Maine, but over the past century she has managed to find her way to the west coast, where she lives in seagrass meadows and floating algae from San Diego to British Columbia. An intrepid adventurer, Amy has lately been spotted in the Netherlands, Portugal, France, and Argentina. In California, Amy is especially fond of eelgrass flowers and seeds that grow close to shore, but because of that she has come into contact with ocean-dumped pharmaceutical wastes, and the exposure has been increasing over time. Concentrations greater than 1 percent waste to water cause coastal dwellers like Amy to stop reproducing and become more susceptible to predators. At 3 percent, she cannot last more than three weeks before she dies. Her tender offspring cannot last a week.

Public water managers have gotten pretty good at detecting and controlling bacteria, viruses, pesticides, petroleum products, strong acids, and some metals. However, until recently, chemicals from prescription drugs and over-the-counter medications have largely gone unmonitored. Sewage treatment plants are simply not designed to remove pharmaceuticals from water. Nor are the facilities that treat water to make it drinkable. That's a scary thought.

The typical medicine cabinet is full of unused and expired drugs, and probably the most common way to dispose of them is to flush the old pills down the toilet. That is what most assisted care facilities do. The toilet and shower are also the routes drugs take to the ocean after passing through our bodies, along with perfume, cologne, skin lotions, and sunscreens. Our bodies metabolize only a fraction of most drugs we swallow, and the rest leaves in excrement or sweat. Harvard Medical School calculates that a single user of testosterone cream can put as much unassimilated hormone into the water as the natural excretions of 300 men.

HOW TO KEEP PHARMACEUTICALS OUT OF THE WATER

- **Limit bulk purchases.** Volume discounts make the price attractive, but big bottles of unused pills create an opportunity for medications to end up in the water.

- **Use drug take-back programs.** Drug take-back programs allow people to drop off their unused medications at central locations, keeping unused drugs from being haphazardly discarded. In the United States, federal law since 2010 makes it easier for those programs to be organized at a local level, so you may have one in your community.

- **Do not flush unused medicines or pour them down the drain.** Some authorities still advise that certain powerful narcotic pain medications should be flushed (unless you can find a drug take-back program that will accept them) because of concerns about accidental overdose or illicit use. It is much better if you don't have a drug take-back program to remove them from their packaging and send them to the sanitary landfill. If you want to be extra careful, crush them first and mix them with coffee grounds so that no person or seagull might scavenge them later.

In the US, one billion tons of annual animal waste from large-scale poultry and livestock farms are laced with hormones and antibiotics fed to animals to make them grow faster and to keep them healthy in close confinement. Often that ends up in waterways. A survey taken around the time Xen was born found medications in 80 percent of the samples from 139 streams, some of it from septic tanks, some of it from farms, and some of it just tossed in the trash somewhere.

Numerous studies have shown that birth control pills and post-menopausal hormone treatments have a feminizing effect on male fish and can alter female-to-male ratios. Intersex fish—creatures with both male and female sex characteristics—have been found in heavily polluted areas. Other research has uncovered popular antidepressant medications concentrated in the brain tissue of fish downstream from manufacturing and wastewater treatment plants.

Whether they were sick or not, marine animals like Amy are being treated with antifungals, antimicrobials, and antibacterials. They are being given medications for pain, fertility, mood, sleeplessness, and neurodegenerative diseases. A platypus living in a contaminated stream in Melbourne is already likely to ingest more than half the recommended adult dose of antidepressants every day, but by the time Xen's children are Xen's age, the volume of pharmaceuticals diffusing into freshwater could increase by two-thirds.

Serotonin drugs cause shore crabs to exhibit "risky behavior." Salmon smolts exposed to benzodiazepines, such as Valium and Xanax, lose any anxiety about the ocean migration and journey downriver too soon—twice as quickly as their unmedicated cousins. They arrive at the sea in an underdeveloped state and before seasonal conditions are favorable and are unlikely to make it to adulthood.

Antidepressants like Prozac slow the learning and memory of cuttlefish, cause marine and freshwater snails to swim away from the rocks they should be clinging to, and make shrimp swim toward light sources, where many predators (and clever shrimp fishers) await them. Amphetamines speed the juvenile development of aquatic insects and cause male aquatic starlings to sing less and lose interest in females.

Rebecca Giggs, a writer from Perth, Australia, says, "That we may, inadvertently, be changing the mental health of wildlife is an unhappy realization, even if it expands what is understood of animals' emotional worlds."

There has been, almost since the beginning of life, a dilute algal soup floating on the surface of the ocean. This soup has nourished herbivorous zooplankton, turtles, and baleen whales. It draws its ingredients from floating or raining nitrogen compounds, phosphates, carbon, and sunlight. If you dip a ladle into this soup, you can see hundreds of small spherical cells. If you ride in a boat through it or swim through it on a clear moonlit night in mid-June, it will glow an iridescent blue from bioluminescence that it produces when disturbed. If you see foam on a beach after a storm, it may well be millions of dead bodies of this bioluminescent algae.

There are sometimes conditions at sea—calm seas, sunlight, sediments—that favor the growth of the algal soup so much that the population of algae explodes to millions per liter of water. It grows so dense that it is dangerous to fish, clogging their gills and mouths. Red algaes produce a "red tide"—and hundreds of tons of floating fish carcasses. As the bloom exhausts its food and dies, the algal cells sink to the bottom, smothering anything living on the seabed.

These overactive algae also tend to concentrate bacteria up the food chain as they are consumed by shellfish, crabs, and baby turtles. Mussels can easily eat over 50 million cells per hour, storing and concentrating the bacteria. Outbreaks of permanent paralysis and other diseases in coastal cities sometimes follow red tides and have been traced to this shellfish toxin. The human communities most vulnerable to these biological hazards are those without sustained monitoring programs and dedicated early warning systems for harmful algal blooms.

As a dilute soup, the algae are life-givers, but as a red tide, they are deadly. A combination of warming oceans, sewage, soils from rivers, and dying fish is providing the conditions to make such tides more frequent. Even as algal blooms increase, the larger forms of ocean life that Xen sees now are getting weaker, and as they weaken, the web of life—not just in the ocean, but also on land—breaks. None of us can survive unless we can repair the web of life.

8

Oxygen and Acid

If you have ever been in a smoke-filled room and wanted nothing more than to go outside and get a breath of fresh air, you can relate to the current experience of your average flounder. It's floundering. Fish don't breathe air, but they still take in oxygen dissolved in the water around them, and if those oxygen levels drop too low, they can strangle.

We don't often think about oxygen in the ocean. And yet, many major extinction events in Earth's history have been associated with oxygen-starved oceans, most often from climate change.

Since the onset of industrialization, widespread burning of fossil fuels, deforestation, and cement production have released more than 1,400 billion metric tons of carbon dioxide (CO_2) into the atmosphere, about half of that since midway through the twentieth century. The ocean absorbed around 27 percent, dissolving it into seawater and making carbonic acid that made the ocean more acidic. Over the last 200 years, seawater has become 30 percent more acidic. At present rates, by 2060, seawater acidity could become 120 percent greater than in pre-industrial times.

"Knudsen salinities" can be used to measure salinity, expressed in units of parts per thousand. Normal seawater is euhaline, in the range of 30 to 35 Knudsens. Metahaline bodies, like the Red Sea, range from 36 to 40. In some places, inland seas can go as high as 300 Knudsens. The saltier a body of water is, the less likely it is to absorb carbon dioxide from the atmosphere and the more likely it is to give it off. This is a critical recovery element at the end of ice ages, when salinity peaks due to ice impoundment of freshwater from rain or snow, causing more CO_2 to off-gas to the atmosphere and positively force the greenhouse effect, rewarming the world.

The degree of salinity in oceans is a driver of the world's ocean circulation, where density changes due to both salinity changes and temperature changes at the surface of the ocean produce changes in buoyancy, which cause sinking and rising of water masses. The sinking and rising exchanges warm water from the surface and equator with cold water from the depths and the poles, and the transfer of heat stabilizes the interior climates of continents. Heat exchange is the engine that drives major ocean currents like the Atlantic Meridional Overturning Circulation (AMOC).

DESALINATED WATER

Most human drinking water, as well as water for industry and agriculture, depends on the rain. The exception is desalination, a process of removing salt from brackish or mineralized water, such as seawater. Until recently, desalination was principally used by ships and submarines, but now, with growing water scarcity due to population growth, urban development (primarily in coastal areas), and climate change, desalination is a rapidly emerging industry, and one that may affect the ocean.

Desalination processes are usually driven by either thermal energy (in the case of distillation) or electrical energy (as with reverse osmosis). The requirement for energy can make desalination a costly or even unfeasible option for many localities. Still, for those with abundant energy, such as Qatar, with 99 percent of its drinking water now coming that way, or for those willing to spend money, such as Cape Town after three years of a devastating drought, it is a welcome choice.

Desalination facilities worldwide now include about 16,000 operational plants with a global capacity of more than 95 million cubic meters per day (m3/d). Seawater accounts for the most significant volume (59 percent), followed by brackish water (21 percent) and other less saline sources. New ocean-water desalination projects are on the rise, including floating desalination plants constructed on ships and offshore structures, which have the advantage of being mobile.

The GivePower Foundation placed solar arrays on 2,650 schools in 17 countries before it began to see that light for reading was not the only problem solar energy can solve. In a small Kenyan village, it saw the devastating effects of brackish water, with children dying of kidney failure and water-related diseases getting worse. GivePower began a new initiative to establish solar-desalinated water plants, delivered in shipping containers, serving up to 35,000 people.

Today, approximately one percent of the world's population is dependent on desalinated water to meet daily needs, but the number could rise to 14 percent by 2025.

As with all new things, we need to look at the entire picture. When the salt is removed, where does it go? Seventy percent of the world's desalination plants are in the Middle East, where, for every one cubic meter of fresh water produced, two cubic meters of salty brine are discharged into the already salty Arabian Sea. The world's 16,000 desalination plants discharge 142 million cubic meters of brine daily—enough in a year to cover an area the size of Florida under a foot of brackish, heavily salted water.

Other environmental concerns associated with desalination include the greenhouse gases and air pollutants that can be generated when natural gas or diesel-electric plants provide the energy required; the capture of marine biota during the intake of feedwater; and the discharge of chemicals used during pre- and post-treatment phases. High salinity and elevated temperature in the effluent can be fatal to marine organisms and impair coastal ecosystems. The high salinity elevates density in comparison to the receiving waters, which can form "brine underflows" that deplete oxygen. Desalination infrastructures can also alter coastlines and how sediments are transported and deposited. They can impair migratory seabirds and nesting sea turtles.

With increasing water demands coupled with water scarcity, desalination is expanding, and so is the brine produced. Recent efforts have focused on ways to treat or use brine to minimize or eliminate the negative impacts. Techniques such as bipolar membrane electrodialysis can recover many scarce metals. Brine has also been used for spirulina cultivation and the irrigation of halophytic forage shrubs and crops in aquaculture and aquaponic systems.

The International Desalination Association, established in 1973, is the world's leading resource for information and research for the global desalination industry, with 4,000 members across 60 countries.

The ocean has absorbed over 90 percent of Earth's additional heat since the 1970s, and this has led to ocean warming and decreasing oxygen. Currents slowing because of ice melt at the poles has meant less mixing in a warmer ocean, particularly in the tropics. That, along with fewer whales and dolphins, may bring fewer nutrients from nutrient-rich deeper waters to the nutrient-poor surface waters, which will slow photosynthesis by algae. With fewer algae, we will have reduced fish, seal, and turtle populations. Those marine species that can swim are shifting toward cooler waters, but many stationary organisms, such as corals, have no way to escape from warmer, deoxygenated, or more acidic waters and are left behind to wither and die.

The warmer a liquid becomes, the less oxygen or any other gas it can hold. Because of surface warming and excess carbon dioxide taking up more space, the ocean has given up an estimated 2 percent, or 77 billion tons, of its oxygen over the past 50 years. As a result, oxygen-minimum zones (OMZs) have expanded by an area about the size of the European Union (4.5 million square kilometers). Anoxic dead zones (devoid of oxygen) have more than quadrupled. For your average fish, this means trying to steer clear of OMZs and dead zones, mainly along the coasts, where corals and mangroves once created optimal conditions. Unfortunately, those corals and mangroves are now full of topsoil, agricultural chemicals, plastics, radioactivity, and deadly algal blooms. As schools of fish try to swim to where they can breathe, they become easy targets for the sophisticated sonar systems and satellite services now affordable to every commercial or sport-fishing boat. Fishers merely glide to the edges of oxygen-starved zones and wait for the catch to come to them, or float to the surface.

Just the upper 3,300 feet (1,006 m) of ocean has lost up to 3.3 percent of its oxygen since 1960. If you are at sea level, the air you breathe is about 20 percent oxygen. Removing 3.3 percent would be like hiking up to 5,000 feet (1,524 m). That is still breathable—about like Mt. Washington, New Hampshire—but consider that the recent 3.3 percent change was just over the past 60 years and is accelerating. There might be a 6–8 percent difference in another 30 years, which would be like climbing Pikes Peak. We can still breathe, but for how much longer? At 10–12 percent change, we would need to bring oxygen tanks to climb any higher. What is a flounder going to do?

Aside from food-web disruptions, animals face various other physiological challenges as their bodies adjust to lower oxygen levels. Some, like jellyfish, are more tolerant of low oxygen than others, but all marine animals will feel the impact of deoxygenation because they all have evolved their oxygen capacity for a reason. Chinese shrimp flip their tails less vigorously to conserve energy in lower oxygen. Some male fishes produce fewer sperm—and the trend does not seem to bounce back in future generations when oxygen levels improve. Sensory functions, such as seeing and hearing, also suffer in an oxygen-depleted ocean. We know this ourselves because when we travel at high altitudes, we can experience reduced night and color vision. Many species rely on reflexive visual cues to avoid predators, so any sight loss would make it more difficult for them to survive.

Healthy ocean phytoplankton plants and green microbes cannot assimilate carbon dioxide or liquid carbon dioxide solutions like the leaves and roots of plants in your garden can. Unlike their land cousins, ocean plants consume bicarbonate (HCO_3), not carbon dioxide (CO_2). So, for every atom of carbon, they need to find one atom of hydrogen and three atoms of oxygen. Over the course of a billion years, phytoplankton developed a few tricks to make this easier. They make an alkaline membrane that slows bicarbonates from leaving the ocean in normal evaporation. That way, they always have plenty of bicarbonate handy. All plants also need nitrogen, sulfur, and halogens, and in the ocean, these fertilizers arrive from dissolved salts and excretions from sea creatures. Metabolic reactions convert arriving nutrients into useful forms for the plants. Those reactions also produce hydroxide ions needed to make bicarbonate.

These ocean gardeners have been so successful in holding bicarbonate and generating basic ions that, left on their own, they would turn the entire ocean alkaline, with a pH value greater than 9 (neutral is 7). What prevents that is a return program for solid carbonate shells and skeletons from coccolithophores, foraminifera, shellfish, and corals. After their owners no longer need them, those discarded shells build up upon the ocean floor. That continuous carbon sequestration keeps the ocean's pH value well within the metabolic optimum for marine life. During some ancient periods when phytoplankton productivity was much higher than now, the ocean's shell makers were so productive that they laid down deep, chalky layers of carbonaceous sediments. The white cliffs of Dover are one example.

Reef-building, carbonate-storing coral colonies require clear water. The tiny coral animal—the polyp—produces most of its food through photosynthesis by algae in its tissues. Sedimentation, which screens out sunlight and kills polyps, is one of the principal reasons coral reefs bleach and die. Desedimentation by seagrass and filter-feeding mollusks keeps the water clear. Seagrass meadows thrive in calm waters where the barrier reef protects them from the surf and wind-driven currents. Just as the coral polyp lives in internal symbiosis with its resident algae, so the coral reef lives in external symbiosis with seagrass.

Because alkaline-producing reactions only happen during daytime when plants are active, the pH of the ocean decreases during the night. In the polar regions during long winter nights, the constant dark creates holes in the plankton net and allows more CO_2 to escape from upwelling deep water without being restrained as bicarbonate to feed the plants. Fortunately, the balance tips back during the long summer days when whales return to polar regions to fertilize the blooms and CO_2 is once more arrested, converted, and sequestered.

As concentrations of carbon dioxide go up in the atmosphere—from the burning of fossil fuels, the production of cement, and changing land from forests and meadows into farms and cities—so does the carbon dioxide arriving, by rainfall, to the sea. When carbon dioxide dissolves in seawater, it produces carbonic acid. If you have ever had a fizzy drink, this is where the fizz came from.

Adding fizzy acid to the ocean is bad news for anything making chalky shells and skeletons. Acid dissolves seashells and corals. It will also affect hard- and soft-shelled creatures like clams, lobsters, shrimp, oysters, sea urchins, mussels, and those little floating plankton that like to build reflective shields—pteropods, coccolithophores, and foraminifera. Coralline algae, a group of calcareous seaweeds, use alkaline minerals to provide the cement that makes coral reefs.

Since the beginning of the Industrial Revolution, ocean acidity has increased much faster than in the past 50 million years. Its pH (for which a lower number means it is more acidic) has gone from 8.2 to 8.1. This might not sound like a big difference, but when oceanic pH reaches 7.9 or lower, coral reefs and calcifying organisms start to die, setting off a cascading collapse of ecosystems.

More than 30 percent of the world's coral reefs have died over the past 30 years, and 90 percent are projected to die by 2050, due to acidification, other forms of pollution, and overfishing. According to the World Resources Institute, coral reefs generate nearly $30 billion annually and sustain 25 percent of marine life and nearly 1 billion people through coastal protection, food security, and income.

National Oceanic and Atmospheric Administration scientist Emily Osborne's research team measured changes in the thickness of seashells and corals. She was able to estimate the ocean's acidity level during the lifetimes of the microbes that built the reefs. She concluded that the waters off the California coast had a 0.21 decline in pH over a 100-year period dating back to 1895, more than double the decline average worldwide. The data revealed an unexpected connection to the Pacific Decadal Oscillation, a warming and cooling cycle involving strong winds that pull warmer surface water onshore or offshore. The warmer, more nutrient dense, and carbon rich the waters, the faster and stronger the acidification.

Since the effect was first noted in the nineteenth century by the great-grandfather of Swedish climate activist Greta Thunberg, Svante Arrhenius, ocean acidity has risen 30 percent. If carbon dioxide emissions are not curtailed, acidity is expected to rise 150 percent by the time our Xen is 38 years old (Greta will then be 47). Carbonic acid at a strong enough concentration will not just make new reefs and shells impossible, but it will also begin to dissolve the existing ones.

There is evidence that this process has already begun in places like the Great Barrier Reef along the coast of Australia. Coral "tree rings" have been counted to show that the weakening of reef cement began about the same time as the first European colonial settlements arrived. In the Mediterranean Sea, corals today lay down about half as much cement as they did at the start of the Industrial Revolution. Without the cement, the coralline algae cannot build reefs.

Cold water holds more carbon dioxide than warm water, but warm waters near the surface—the top 300 feet (91.4 m)—are the most acidic since the CO_2 is mostly coming from the surface. Recent studies suggest the horizon of acidity is going deeper by 3 to 6 feet (0.9 to 1.8 m) per year, and as that horizon moves, coral reefs crumble and shellfish scurry.

Many whales will swim thousands of miles every year to the cold waters near the North and South Poles to consume the banquet provided by marine algae there. They gorge themselves on wingfoot snails, which are the size of a lentil and swim with a foot that expands into a pair of delicate wings. Wingfoots have been called the potato chips of the sea. If you have ever eaten salmon, cod, pollock, Arctic gar, or other species from polar seas, you have tasted them secondhand.

Parts of the ocean off northern Canada have already become too acidic for wingfoots to make shells. The Southern Ocean around Antarctica is predicted to be too sour by 2030, the Bering Sea by 2100. During Xen's lifetime, he may witness the end of these little potato chips and, along with them, all the sea animals that depend upon them.

The effects on filter feeders in harbors and estuaries will also be severe. There was a time when mussels and oysters were so abundant in the harbors of Boston, New York, and Charleston that the waters would be completely filtered and freshened every day. To accomplish the same amount of filtration today would take months or years because the mussels and oysters were either all fished out or never allowed to recover fully. Acidity threatens the populations of filter-feeding shellfish that remain, and so threatens clean water.

Ocean acidification damages the phytoplankton ecosystem by binding more bicarbonate and reducing OH ion productivity. If, on the other hand, phytoplankton were allowed to thrive in the oceans, more bicarbonate would be produced, alkalinity would be restored, there would be less CO_2 and more oxygen at the surface, and atmospheric CO_2 could continue to be absorbed by the ocean. As ocean waters grow more acidic, there is also a threat to sequestration on the seafloor because more of the discarded plankton skeletons are dissolved, releasing their carbon to the water column and, eventually, some of it back into the atmosphere.

Not only are we not building new white cliffs of Dover in the seabed, but we also are dissolving any that might already be down there.

As we find ways to assist the ocean's gardeners and reverse some of these trends by helping plankton on the surface and corals on the floor, there is one other wild card we have to contend with—the effects of

plastic. Plastic leachate is produced when sunlight, seawater, and bacteria begin to break down floating bits of plastic and make some synthetic polymers soluble enough to be absorbed by plants. In *Prochlorococcus*, the most abundant photosynthetic organism in the ocean, or, for that matter, on Earth, plastic leachate impairs growth and reduces its ability to photosynthesize. More ominously, plastic causes what are called global errors in transcription—the way the organism codes new cells to make future cells. When there are transcription changes, which can also occur with exposure to radioactivity, entire generations of *Prochlorococcus* can be born defective, diseased, and effectively sterile.

If we are concerned, and we should be, there are things we can do right now to reverse this. The first thing is to stop nutrient runoff, plastics, and acid discharges from entering rivers and estuaries. Some cities are using large biochar filters to span urban outfalls, intercepting nitrogen and particulates. Biochar is good at capturing, adhering to, and storing many different toxins.

While some long-term changes cannot be avoided, others are reversible if we act soon. If nations muster the will, such as through meetings like the United Nations summits, to cut emissions from fossil fuels to zero over the next few decades, that might give our ocean friends a breather.

Increased saturation with carbon dioxide and microplastic impairs the ability of fish to breathe, almost as severely as an absence of oxygen. As fish take in water through their gills, the CO_2 diffuses into cells and displaces the oxygen molecule, O_2, suppressing respiration and the circulation of oxygen to the heart. It is as though they were breathing smoke from a wildfire. Stressed this way, they have to devote all their energy just to breathing and so move lethargically, eat less, grow more slowly, and do not reproduce.

They would not want to wish such a world on their children.

PART FOUR

C H A P T E R

The Whale's Tale

It should not be surprising to learn that since 70 percent of the world is covered by water (the Pacific Ocean is larger than all Earth's continents put together), most of the living organisms on our planet live there, organized into interactive communities called ecosystems. Over the history of our planet, some of the most unique creatures Earth has ever witnessed have lived nowhere else but in the ocean.

Ocean habitats vary a great deal. Near polar regions, surface water temperatures go below freezing—about 28°F (-2°C)—but don't freeze because they are always moving. Warmer waters near the equator reach temperatures of 97°F (36°C), about the same temperature as your blood. The average surface temperature of all Earth's oceans is 63°F (17°C), but that is just at the surface, where the sun shines. Temperature decreases as we go deeper, beyond where sunlight can penetrate. The dark, cold zone known as the *thermocline* begins between 300 feet (90 m) and 1,300 feet (400 m) below the ocean's surface. All of these different areas—surface waters, thermocline, polar regions, tropical regions, deep water—create unique habitats for creatures that have adapted to them.

These organisms depend on each other for food—the energy they need to survive. Those close to the surface have a primary source—the sun. Those living in the thermocline cannot get this and so must either rise toward the surface to eat or rely on other means to bring food down to them.

The plant-like organisms found at the surface are producers of energy for other organisms in the ecosystem. They are called *producers* because they make their food by absorbing light from the sun and converting it, by photosynthesis, into sugars that can feed cellular growth. Looking at producer organisms that we can see with the naked eye, such as phytoplankton, algae, and seaweed, does not reveal the full picture, however. Cooperating with these single-cell and multicell producers are microscopic creatures, such as bacteria, fungi, and viruses, that help them perform their tasks, including photosynthesis. These organisms may be so small that they fit within the empty space in the center of the DNA double helix, assisting in the exchange of genetic material or flipping epigenetic switches on and off.

For many years, ocean sciences did not consider these microbes very important. Now we know that they are the fabric that holds together the web of life, on land as well as in the sea. They are many times more numerous than the more complex life forms. There are 15 times more viruses, for instance, in the ocean than all other types of marine life combined. To count them, you would have to use 30 zeros. If you laid those viruses end to end, it would make a string two hundred times finer than the most delicate spider string, and it would extend out 200 *million* light-years, passing along the way 40 galaxies as massive as our own Milky Way.

Sea creatures that rely directly on producers for their food are called primary consumers. When a primary consumer feeds on a producer, the energy made by the producer (in a cooperative effort with its microbiome) is transferred up the food chain. Corals, lobsters, and clams are primary consumers, and so are whales, sea turtles, and sharks. You can think of this as an energy pyramid, as large fish eat smaller fish. Even decomposers like worms and bacteria that break down the dead and decaying material when producers and consumers die are a part of this pyramid, feasting from all levels, top to bottom.

The health of this pyramid food system is essential for the health of the ocean, because all life in the sea is dependent on all other life, just as life on land is dependent on the health of the ocean, as we shall see.

Blue-green algae, living at the base of the food chain, has been around for most of the life of our planet—some 3.5 billion years, according to the fossil record. Their proper scientific name is *cyanobacteria*, which means bacteria that do not require oxygen. But they actually are responsible for our oxygen atmosphere, because for billions of years they breathed in gases that we cannot—such as carbon dioxide and nitrous oxide—and converted those to elements such as carbon, nitrogen, and oxygen, the building blocks of life. As blue-green algae inhaled toxic gases formed at the birth of our planet and exhaled oxygen, they built the atmosphere we breathe today.

Blue-green algae contain chlorophyll, which they and their symbiotic microbiome use to capture sunlight and store its energy as green cells, just like plants. There are more than 800 ocean species of blue-green algae, but many more, nearly 8,000, live in freshwater. The algae cells group themselves in stringy clusters of filaments that network together to make colonies. These colonies may collect the shells of dead diatoms and floating microplastics for shelter from wave action, or for insulation from temperature changes.

There are other tiny producers called plankton that move around the world's oceans on tides and currents. Some are self-mobile, and we call them "animals," or zooplankton. Others grow on sunlight and minerals drawn from the sea, and we call them "plants," or phytoplankton. Most plankton are too small to be seen, other than what appears to be a change in color of the water, but they are the most common producers in these ecosystems and form the foundation of the food chain for all other ocean creatures.

The word "plankton" really describes only how these organisms move around, and many organisms we know as larger sea creatures—crabs, sea stars, seahorses, and most fish—begin life as plankton until they grow large enough to be recognized as something else. Many types of jellyfish are considered zooplankton because they travel on waves and currents. Other jellyfish can propel themselves very deep into the ocean with rings of muscles that flex in unison. Some, like the Nomura jellyfish, can be 6.6 feet (2 m) and 20 feet (6 m). There are more than 3,000 kinds of jellyfish, and none of them have what we might call a brain. They have a basic nervous system that senses light and nutrients in the water. Most are transparent and bell-shaped, with long tentacles that contain venom, which they use to sting their prey. This makes them secondary consumers—those that consume primary consumers.

Diatoms are tiny organisms related to plants that have a hard shell that sinks when they die. Diatoms do not have roots, stems, or leaves, but they make their shells of shiny silica, the necessary ingredient of glass. Of the more than 10,000 diatom species, the largest can be seen without a microscope. Like plants, they contain chlorophyll, and like other types of plankton, they form colonies that clump together. These organisms serve the vital function of reflecting vast amounts of sunlight into space, maintaining an even temperature at the surface of the Earth.

There are also organisms called coccolithophores that form shells like those of diatoms by taking calcium carbonates from seawater. Unlike diatoms, the shells are not based upon silica but rather upon carbon. This is an important way—perhaps the most important way—that carbon is removed from Earth's atmosphere and stored back in the land.

The coccolithophores drape their ball-shaped bodies in a bright ceramic-carbon armor and, in seasons when the days are longer and the nutrients arrive on upwelling currents, form massive blooms that turn the ocean surface milky white. These swirling white islands of coconut-colored coccolithophores are so large they can be photographed from space. Each bloom gathers millions of tons of calcium carbonate from the water. When conditions change and the coccolithophores die, they and their hard shells drop to the ocean floor, entombing all that carbon in soon-to-be limestone sediments. Aiding this process are the ocean viruses that proliferate when the plankton begin to run out of food and liberate nutrients to be used again by other plants and animals.

The ocean is responsible for removing 90 percent of all fossil fuel pollutants that enter the atmosphere every year. When carbon dioxide dissolves in seawater, it forms carbonic acid. If it were not for the more than 10,000 species of plankton absorbing the carbon from that acid and combining it with sea salt to make calcium carbonate for outer shells, the oceans would have oversaturated with carbon dioxide and stopped removing it from the atmosphere long ago, making the land surface too hot to be livable.

Phytoplankton—those little green plants growing on the surface of freshwater and saltwater—are highly efficient carbon sequestrators. Worldwide, their "biological carbon pump" transfers about 10 gigatons of carbon (GtC), or 37 $GtCO_2$, from the atmosphere to the deep ocean each year. Total human activity, from fossil-fuel burning to soybean farming, adds about the same. Human civilization and phytoplankton are a matched respiratory cycle. The International Monetary Fund recently observed:

> These microscopic creatures not only contribute at least 50 percent of all oxygen to our atmosphere, they do so by capturing about 37 billion metric tons of CO_2, an estimated 40 percent of all CO_2 produced. To put things in perspective, we calculate that this is equivalent to the amount of CO_2 captured by 1.70 trillion trees—four Amazon forests' worth.

There are new threats to plankton that were not present in earlier centuries and millennia. The amount of carbon dioxide trying to enter the ocean every time it rains is overloading the oceans with carbonic acid, which is lowering the pH and raising acidity. This change in pH alters the availability of some nutrients, such as iron, that phytoplankton need to thrive. As ocean acidification increases, the areas available for plankton growth shrink, which means the basis for the ocean food chain shrinks, the carbon dioxide going into carbon shells diminishes, and the food the oceans produce for us is reduced.

During Xen's lifetime, open ocean surface pH is projected to decrease by around 0.3 pH units. That acidity is past a shell-stability threshold. In polar and subpolar oceans, it could be impossible to build corals or make shells by as early as 2081.

SWIMMING WITH THE SHARKS

One of the most memorable thrills of my life was my first swim with Atlantic whale sharks off of México, in the Yucatan Channel. My guide on the boat deck motioned to me—I was treading water and scanning the horizon for any dorsal fins—to "dive, dive!" and I did, just in time to see a shark some 30 feet (9 m) long, with an open mouth nearly 5 feet (1.5 m) wide, converging directly on my position, at speed.

I dodged to my left as fast as I could, not hazarding the time it would take to bring my camera up and record the moment. By the time I positioned my small point-and-shoot, the shark was past me, devouring plankton to the starboard of the boat. These giant sharks are herbivores, so he or she was probably trying to avoid me as much as I was trying to avoid him or her. Here is the picture I took:

(continued)

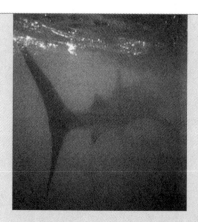

Whale sharks, currently on the Red List of Threatened Species, first appeared in the geological record, along with manta rays, at the Eocene-Oligocene boundary about 34 million years ago. That was when one of the largest extinctions of marine invertebrates and mammals in Europe and Asia, likely triggered by volcanism, dropped Earth's temperature some 59°F (15°C) over a few hundred thousand years, isolating a warm water niche near the equator at a time when the continents were much closer together. Whale sharks are 100 times older than our current evolutionary form as large-brained, upright humans.

Whale sharks, called "dominoes" in the Caribbean because of their white spots, can live up to 130 years and grow 60 feet (18 m) long. They are well adapted to warming oceans and only rarely found in waters below 70°F (21°C). Their annual migration routes in both the Eastern and Western hemispheres track plankton blooms in warm waters.

Without these plankton feeders, and those that feed on them, more plankton would decompose close to the surface, on their way to the depths, or in shallow sediments, returning their carbon to the atmosphere as carbon dioxide or methane, which are climate-changing greenhouse gases. There would also be a substantial loss of the nutrient flows that fertilize plankton blooms, significantly reducing their extent.

With their vegan diet, whale sharks are giant carbon sequestrators, peacefully migrating north and south to browse on green floating meadows.

Toothed whales, as well as baleen whales, are descendants of land-dwelling mammals of the *artiodactyl* order (even-toed ungulates). Unlike the whale sharks, which use baleen to filter large volumes of water to feed, modern toothed cetaceans, including dolphins and porpoises, track prey using sonar. They echolocate by emitting a series of clicks at various frequencies from their melon-shaped foreheads that reflect off objects and are retrieved through the lower jaw. *Odontoceti* (toothed cetaceans) also use enhanced fat synthesis to store calories in insulating blubber, giving them a wider range of habitats and a greater depth of dive. Many baleen whales and sharks never mastered that blubber trick, and that keeps them in warmer surface waters.

Until the industrial era, phytoplankton were the thermostat that moderated human- or volcano-induced warming to maintain a comfortable "Goldilocks Zone" balance between too-cold Iceball Earth and too-warm Hothouse Earth. Whenever river, lake, or ocean surfaces began to warm, tropical growths of plankton would speed up, converting carbon dioxide into oxygen and carbon. The oxygen would rise to the atmosphere, freshening the air, and the carbon would be eaten by fish, whale sharks, or sea turtles or descend to the depths to be decomposed by worms and sea urchins and be interred into the sediments.

Even small changes in the growth of phytoplankton may affect atmospheric carbon dioxide concentrations, which feed back to global surface temperatures. A huge change would occur if there were any significant loss of plankton-feeders.

When a baleen whale scoops up plankton or a toothed cetacean rises to breathe, it also defecates. The nitrogen, potassium, and other important minerals in their excretions are fertilizer for the plankton bloom. Take away the whales, and you remove this critical source of nutrients. When a whale, turtle, shark, or dolphin dies, it sinks to the ocean floor to be eaten or decompose. If the floor is deep enough, its carbon, as methane, is trapped and cannot rise to the surface and return to the atmosphere.

Economists at the International Monetary Fund have calculated the value of a whale at $2 million each due to the important role they have in reducing greenhouse gases. A blue whale, for example, can take nearly 30 metric tons of CO_2 out of the atmosphere each year, for centuries, compared to around 33 pounds (15 kg) that is captured by a tree each year.

At a minimum, even a 1 percent increase in phytoplankton productivity thanks to whale activity would capture hundreds of millions of tons of additional CO_2 a year, equivalent to the sudden appearance of 2 billion mature trees. Imagine the impact over the average lifespan of a whale, more than 60 years.

Our conservative estimates put the value of the average great whale, based on its various activities, at more than $2 million, and easily over $1 trillion for the current stock of great whales.

We once had 5 million whales in all the world's oceans. Today there are about 1.3 million. By the 1930s, we were killing 50,000 per year, mostly for outdoor lighting and axle grease. Phytoplankton, which generate 50 percent of all oxygen and capture 40 percent of CO_2 produced, rely on whales to provide nutrients and maintain their numbers. If the number of whales was restored to around 5 million, this would significantly increase phytoplankton. Even a 1 percent increase in phytoplankton would capture millions of tons of additional CO_2, say the economists.

We estimate that, if whales were allowed to return to their pre-whaling numbers—capturing 1.7 billion tons of CO_2 annually—it would be worth about $13 per person a year to subsidize these whales' CO_2 sequestration efforts.

Since the role of whales is irreplaceable in mitigating and building resilience to climate change, their survival should be integrated into the objectives of the 190 countries that in 2015 signed the Paris Agreement for combating climate risk.

Since, thanks in no small measure to Greenpeace and Sea Shepherd, whaling is largely curtailed today, having been reduced to renegade Icelandic and Japanese research vessels and coastal dolphin hunts, the populations should be recovering, right? Wrong.

Enter plastics. Globally, at least 23 percent of marine mammal species, 36 percent of seabird species, and 86 percent of sea turtle species are now threatened by plastic debris. So far, over 270 species, including turtles, fish, seabirds, and mammals, have been observed to have experienced impaired movement, starvation, or death from exposure to plastics. However, studies in the northern Gulf of California found that as few as 1 in every 50 to 250 carcasses (0–6.2 percent) is recovered from cetacean deaths at sea. Some 300 other species of marine fauna are thought to also be in jeopardy. Among marine mammals, the ingestion of plastic has been documented in 48 cetacean species (56 percent), with rates of ingestion as high as 31 percent in some populations.

Plastics are forever. Approximately 10,000 shipping containers plummet off cargo ships into the ocean each year. A shipping crate carrying 28,000 plastic ducks was lost at sea between Hong Kong and the United States in the Pacific Ocean over 20 years ago, but at least 2,000 of the ducks are believed to be still circulating, while others have washed ashore in Hawaii, Alaska, South America, Australia, and the Pacific Northwest.

We are increasing not only the amount of plastic in the environment, including microplastics prone to ingestion by baleen filter-feeders, but also the rate by which that amount is growing. We are just at the upward turning junction of the J-curve. In the 1960s, five million tons of plastic were manufactured worldwide each year. By the time Xen turns 38, 30 years from now, there could be 1,800 million. It's exponential. Almost one-third of plastic produced is used to manufacture single-use "disposable" plastics, such as coffee cup lids, stirrers, and straws. Each year, we design and sell more "disposable" products using materials that float and last forever.

What is being "disposed of" with every straw, stirrer, bathtub duck, and Halloween costume are the whales. And with the whales go the ocean's thermostat. The International Monetary Fund has offered both a prescription and a warning:

> Healthy whale populations imply healthy marine life, including fish, seabirds, and an overall vibrant system that recycles nutrients between oceans and land, improving life in both places. The "earth-tech" strategy of supporting whales' return to their previous abundance in the oceans would significantly benefit not only life in the oceans but also life on land, including our own.
>
> With the consequences of climate change here and now, there is no time to lose in identifying and implementing new methods to prevent or reverse harm to the global ecosystem. This is especially true when it comes to improving the protection of whales so that their populations can grow more quickly. Unless new steps are taken, we estimate it would take over 30 years just to double the number of current whales, and several generations to return them to their pre-whaling numbers. Society and our own survival can't afford to wait this long.

We have to save the whales. There is no escaping that. But to save the whales, we must also save the turtles, and the gulls, and those little green plants that bob on the waves. To do all that, we'll need to banish non-biodegradable plastics from our lives.

10

Ice Water

Most seals live in the Arctic and Antarctic. The Weddell seal can dive 2,000 feet (610 m) and stay submerged for an hour. Elephant seals can grow to 20 feet (6 m) and weigh more than a ton. Walruses, which can be almost as large, use their whiskers to burrow around the seafloor looking for clams and mussels.

Seals, walruses, polar bears, and other inhabitants of polar regions are showing signs of stress. As the ice breaks up earlier and re-forms later, that affects survival rates of both young and old. As their hunting range declines, the seals and bears are also thinner and weaker, which makes them more prone to sickness or drowning. In some Arctic regions, including the "Last Ice Area" (where summer sea ice lasts longest), polar bears appear to be doing well. However, there are several gray areas on the map—literally and figuratively—where population declines have been seen or scientists just don't have enough information to know what's happening yet.

The fate of penguins, seals, walruses, and bears, unfortunately, depends on the future of the sea ice. As the ice retreats, we see aerial photos of strandings of animals on isolated floes, drifting far from land and too far from other large floes to swim in search of food or safety.

Polar bears cannot swim as far as seals, so they are more likely to die if stranded. Seals can rest on the remaining ice as it floats around and dive for food, but when that ice goes, they, too, no longer have any home.

WHAT DOES A WEDDELL SEAL SOUND LIKE?

Hear a clip of the Weddell seal and many other marine mammals at Discovery of Sound in the Sea: https://dosits.org/galleries/audio-gallery/marine-mammals/pinnipeds/weddell-seal

Of course, Arctic seals and bears are not alone in their plight. The changes in snow cover, lake and river ice, and permafrost, as well as glacier retreat and snow cover loss in the Himalayas, Andes, and elsewhere, have disrupted food and water availability, herding, hunting, fishing, farming, and gathering—harming the livelihoods and cultural identity of a wide diversity of people.

More than 30 years ago, over 1,000 scientists warned the United Nations that unless we did something to reverse fossil fuel use, this ice-melting process could get out of control within 20 years. We didn't, and it did. Positive reinforcing feedback has begun. Now, while we might still be able to slow the process by reducing the concentration of carbon dioxide to below 350 parts per million (such as by reforesting land), we will continue to lose ice from Greenland, Antarctica, and other places to the ocean, and we will lose the reflection of sunlight they provided in the past.

Between 1979 and 2016, the light energy absorption of the world ocean went from 0.21 to 0.71 Watts per square meter (W/m2), which was a huge leap. It had the same effect on global warming as releasing one trillion tons of carbon dioxide, or about 40 percent of all human emissions since the beginning of agriculture. If (or rather when) all the sea ice has melted, the summer months will see a world ocean absorption of 2.24 W/m2, more than 10 times the heat absorption the ocean had in 1979. Since most of our planet's surface is ocean, that means that the Earth will absorb nearly 10 times the heat from the sun as it did in 1979.

However, warmer temperatures at the equator also mean more evaporation, and this leads to more cloudy skies. White clouds reflect light back to space. So does much of the airborne pollution that comes from cities and industries, a phenomenon known as the "global dimming" effect. So far, this increasing cloudiness elsewhere has more than compensated for the Arctic albedo loss.

What will warming polar oceans be like? One change is being seen in the top frozen layer of the seafloor at high latitudes. Over the past 15,000 years, plankton, fish, and penguins dying and falling to the polar seafloor have been in waters so cold that even cold-tolerant microbes could not easily digest them. The methane, carbon dioxide, and other greenhouse gases released by their slow decomposition froze and remained in the shallow sediments. Called "clathrates," these "cages" of frozen molecules, and other organic deposits trapped by land movements over eons and ice ages, contain vast stores of greenhouse gases that, if warmed suddenly, could rise to the surface and be released into the atmosphere. Current estimates are that permafrost zones, both on land and in the sea, contain almost twice the carbon in the atmosphere.

Widespread permafrost thaw—from 24 to 69 percent—is projected for this century. That equates to tens to hundreds of billions of tons of CO_2 and methane. As permafrost melts in the Arctic circle and

WHAT COLOR IS THE OCEAN?

When full-spectrum sunlight, which we see as white light, hits an object, the object absorbs some of the rays and reflects others. A banana reflects yellow light. A grape reflects green or purple. Any object that absorbs all light waves appears black.

What color is the ocean? Large bodies of water absorb longer (red) wavelengths and shorter (violet) wavelengths of sunlight and reflect the blue we see. Water molecules bouncing off ocean waves scatter the blue light all around, which makes the water appear bluer. Dust fertilizing the water to create algal blooms makes the surface reflect green. Soil and sand entering from land combine to make water reflect brown. Shells on diatoms and floating ice reflect white.

A so-called "blue ocean event" refers to an ice-free Arctic when the ice melts in summer, leaving just the open ocean. Objects that are silver or white, like ice, bounce light, while dark objects, like the blue-green ocean, absorb it. When polar ice melts, it transforms the surface of the oceans from a mirror to a dark sponge. While a mirror would bounce away sunlight and leave its surface cool, a dark sponge absorbs it and warms. In this same way, when a warming ocean and atmosphere melt ice from the surface of the Arctic, Greenland, mountain glaciers, or Antarctica, they cause millions of square miles of the Earth's surface to switch from heat reflecting to heat absorbing. This speeds up the melting process, which increases the heat, which speeds the melting. Scientists call this "positive reinforcing feedback."

beyond, in Canada, Alaska, Siberia, and Lapland, it is already releasing vast stores of carbon to the atmosphere. That warms the air, which warms the land, which melts more permafrost, and so forth, in a positive reinforcing feedback loop.

The North Pole and the South Pole are both covered with ice, but they are quite different kinds of ice. In the South Pole, the ice is formed by snow falling on mountains over the landmass of Antarctica. At the North Pole, the ice is formed when water freezes on the Arctic Sea, and snow adds to its thickness. When ice melts in a glass of water, it does not cause the water in the glass to rise because the volume of the floating ice was already determining the level of the water. When ice melts at the North Pole, it is the same—sea level does not change. Ice melting at Greenland or the South Pole is different. Since that is water stored on land, when the meltwater reaches the ocean, the ocean rises.

THE DOOMSDAY GLACIER

The 74,000-square-mile Thwaites Glacier, flowing into Pine Island Bay in the Amundsen Sea of Antarctica, is sometimes referred to as the "Doomsday Glacier." It rests on dry land and is held back from the ocean by a rocky ridge 2,300 feet (700 m) tall, but now the ice is retreating from that ridge, and already 600 billion tons of ice have melted. Each year the thawing of Thwaites contributes about 4 percent of global sea-level rise. It no longer risks sliding into the sea, but it contains a mass of water roughly the size of Florida or Great Britain, and, entirely melted, Thwaites would raise global sea level by about 25 inches (65 cm).

As climate change has accelerated, so has the melting of Thwaites. At the beginning of 2020, researchers from the International Twaites Glacier Collaboration (ITGC) took measurements to develop scenarios for the future of the glacier and to predict the time frame for a possible collapse. The erosion of the glacier by warmed ocean water seems to be stronger than expected, carving a cavity in the underside of the glacier two-thirds the size of Manhattan. In January 2020, the researchers measured the temperature of the water flowing through the cavity at more than two degrees above freezing.

The loss of Thwaites ice to the ocean is no longer a question of "if," but "when?"

Ice floating on the surface of the ocean changes something else when it melts. It reduces Earth's ability to reflect sunlight. Silvery-white surfaces like ice reflect light to space, especially the low-angle light at the poles. Dark surfaces like the blue-green ocean absorb light and store it as heat.

Since satellite observations began in 1979, the extent of Arctic ice in summer months has been shrinking, although it varies year to year. The 13 lowest sea ice extents in the satellite record have all occurred in the last 13 years, from 2007 to 2019. Moreover, sea ice cover has been changing from a robust and thick ice mass in the 1980s to a younger, thinner, more fragile ice in recent years. First-year ice is more likely to melt than older ice, so the record tells us that blue ocean events are more likely during the long summer days of the midnight sun.

Any time we get a blue ocean event, the heat increase to the ocean is the same as adding 25 years' worth of human greenhouse gas emissions in a single season. Even though we have not yet reached that point, the Arctic is warming at a rate twice as fast as the rest of the planet. The Arctic Ocean is already warmer than it has been in 800,000 years.

While the changes in the far north and south may be bad news for many polar residents, it does not mean that life will cease there. On the contrary, by the time Xen is in his 20s and 30s, there will be an undersea march toward the poles. A blue ocean at the North Pole means that previously unheard-of fish migrations now become possible. In 2010, fishers on the coast of Palestine rubbed their eyes in astonishment as a great Pacific gray whale breached before their boat in the Mediterranean. Gray whales have not been seen in the Atlantic since New England whalers killed them off in the eighteenth century.

This was not an entirely new experience in the eastern Mediterranean. Following the completion of the Suez Canal in 1869, eager invaders began flowing into the cool waters of the Mediterranean from the warm coral reefs in the Red Sea. Those included rabbitfish, lizardfish, goatfish, and a variety of prawns, launching productive fisheries for the Ottoman Empire and later in post-Ottoman Egypt, Jordan, Israel, Lebanon, Syria, Cyprus, and Turkey.

Might the same happen to the cold waters of the Arctic and Antarctic? If history is a guide to the future, the answer is likely yes.

11

Sea Rise

The Glass Window Bridge of North Eleuthera is one of the island's more popular attractions. Once a naturally formed bridge of rock, it was destroyed in a hurricane and has been replaced by a human-made version, presently in need of repair. From the bridge, a narrow span uniting the long, thin island, you can see the dark-blue Atlantic churning away to the east and the calm turquoise waters of the Caribbean Sea to the west.

Enormous, mysterious boulders are perched on a cliff there, overlooking the Window Bridge. The stones, we now know after extensive research by many geological experts, were dredged from the ocean floor and left on the cliff like Easter Island statuary in a great storm event early in the past interglacial warming epoch. The largest weighs about 2,300 tons.

It is hard to imagine that the air we exhale lifted these huge boulders, but indirectly, it did. At the end of the Eemian epoch, before the last great Ice Age 115 to 120 thousand years ago, CO_2 in the atmosphere warmed the oceans and created superstorms that pushed these boulders up from the seafloor and lifted them to the cliff top.

Sea level rise relative to coastlines is not gradual. Storms are how changes occur because, in their scale and violence, they carve new landscapes that remain altered when the events subside. New bays are formed, old ones are filled in, barrier islands are erased, peninsulas split into islands, and cliffs are carved from windward or seaward slopes.

About 90 percent of all ocean life prefers the shallower depths of the continental shelves—those relatively smooth and even terrains just off the coasts of continents. In Earth's history, these shelves may at one time have been dry land, as the oceans rose and fell. During the last Ice Age, 12,000 years ago, so much water was locked up in ice that the sea level was 125 meters below today's, and vast expanses of continental shelves were dry land. Then, as the Ice Age ended, the sea level rose again. After 10,000 years, ice is still melting. It melted only very slowly for many thousands of years, but in the past few centuries, the rate has been speeding up, as steam power and fossil fuels have heated the globe.

About 70 percent of Earth's freshwater is stored as ice in places like Greenland, Antarctica, and the Himalayan mountains. Ice loss from Greenland is doubling every decade. At the other end of the world, in Antarctica, it is tripling. Melting glaciers from Alaska to Peru are sending rivers of water to the ocean.

"The future is heading far beyond the range of anything we've seen observed in the scientific instrumental record of the last 150 years," says Chris Turney, Professor of Earth and Climate Science at UNSW Sydney. "We have to look further into the past if we're going to manage future changes."

During the previous interglacial, 120,000 years ago, global mean sea levels were between 19.6 feet (6 m) and 29.5 feet (9 m) higher than the present day. Some scientists suspect it could have reached 33 feet (11 m) above where the ocean is now. A more important discovery is that it took less than 2°C of ocean warming to occur. "Not only did we lose a lot of the West Antarctic Ice Sheet, but this happened very early during the last interglacial," says Turney. It seems this very large body of ice is also one of the most fragile.

Greenland ice is melting seven times faster than it was 30 years ago, losing 370 billion tons to the sea every year. Since 1992, six million people living along the coasts of the world have lost their homes from the water added to the ocean as Greenland melts. Another 250 million live barely 2 inches (5 cm) above the waterline. If the Greenland ice sheet were to melt completely—a process that would likely require a few centuries—sea level would rise approximately 24 feet (7.4 m).

OUR SENTINELS

At the end of 2019, the European Space Agency launched a new weather satellite called Sentinel-6A. Every 10 days, it completes one orbit of the Earth, looking at the sea-surface height. Scientists use the data it collects to predict the timing and intensity of coming storms and heat waves.

"If you think about it, global sea-level rise means that 70 percent of Earth's surface is getting taller—70 percent of the planet is changing its shape and growing," said Josh Willis, the mission's project scientist at NASA's Jet Propulsion Laboratory in Pasadena, California. "So, it's the whole planet changing. That's what we're really measuring."

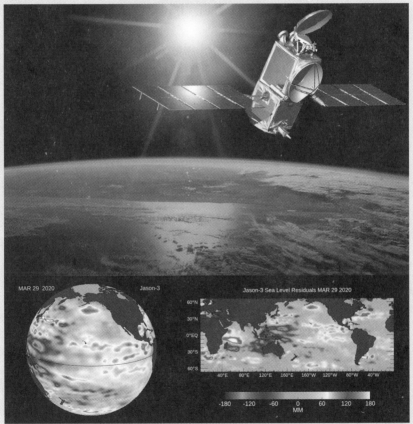

The Jason-2/OSTM satellite provided insights into ocean currents and sea-level rise with tangible benefits to marine forecasting, meteorology, and understanding of climate change. Its successor, Jason-3, is continuing these observations.

Sea-level rise, which is usually given as some global number, is not the same everywhere. Regional differences can make it a third higher or lower than the worldwide average. These differences come from rates of ice melt, variations in currents, expansion due to water heating near the equator, and rapid vertical land movement from activities like building construction or pumping groundwater.

Since Xen was born in 2012, oceans on average have risen less than two inches. From 1900 to 1930, the rate of rise was 0.6 mm/year (two-hundredths of an inch). From 1930 to 1992, the average increase was 1.4 mm/year, more than twice as fast. From 1993 to 2015, the rise was 3.3 mm/year (over one-tenth of an inch). On average, the twentieth century raised the ocean nine inches, 60 percent from ice melt and 40 percent from thermal expansion.

Of 32 tide-gauge stations in locations along the US coastline, 25 showed a clear acceleration in sea-level rise in 2019. The gathering speed of sea-level rise is evident even within the space of a year, with water levels at the 25 sites rising faster in 2019 than in 2018. The highest rate of sea-level rise was recorded along the Gulf of Mexico shoreline, with Grand Isle, Louisiana, experiencing a 7.93 mm annual increase, more than double the global average. The Texas locations of Galveston and Rockport had the next largest sea-level rise increases.

Generally speaking, the sea is rising faster on the US east and Gulf coasts compared with the US west coast, partially because the land on the eastern seaboard is gradually sinking.

What happens when an ice mass melts is that sea level falls closest to the melting ice. That is counterintuitive, but it is because of diminished gravity and the uplifting crust of the Earth once the weight of ice is removed. Sea level begins to rise only when you get 2,000 kilometers (about 1,250 miles) away from the ice sheet. If all the Greenland ice were to disappear tomorrow, the sea level from eastern Canada to Norway would fall. It might have a 30- to 50-meter drop at the shore of Greenland. At the same time, sea level in most of the Southern Hemisphere would increase by about 30 percent more than the global average.

At the other end of the world, if the Antarctic melts, sea levels would fall in Argentina, South Africa, New Zealand, and Australia. They would rise in the Northern Hemisphere, such as along the east and west coasts of North America. The Dutch won't have to raise their dikes to deal with ice melt from Greenland, but they will have to raise them when the ice melts on Antarctica.

The US's National Oceanic and Atmospheric Administration (NOAA) warned in 2019 that if greenhouse gas emissions are not constrained, there may be a worst-case scenario of as much as an 8.2-foot (2.5-meter) sea-level increase this century in some locations.

We're currently in an interglacial—a warm period between glacial cycles. If humans weren't warming the climate, Earth might be slowly returning to another ice age. These changes in Earth's climate happen at regular intervals because of the eccentric orbit and tilt of the planet toward or away from the sun. During ice ages, the weight of ice at the poles squishes the planet and flattens it a little bit. Now, as the ice melts, the weight is lifting, and ice is changing to water, which spreads the load more equally toward the equator, so Earth is becoming more spherical again. Just like when ballerinas or figure skaters put out their arms to slow their spin, this change in the shape of the Earth slows its rotation. Earth's rotation is once more slowing. Days are getting longer.

There are other effects that we can also see happening that affect sea level. Tides crash into the shoreline, and each time they do, they dissipate energy and that slows Earth's rotation. There is a subtle coupling between the core of the Earth, which is iron, and the rocky part of the Earth, the mantle, that acts to change the rotation rate, and as that spin changes, so does the Moon's gravitational pull on tides. Because of tidal dissipation, core-mantle coupling, and the periodic ice ages, we have had a small slowing of Earth's rotation over the last 2,500 years—about 60 milliseconds.

Over the past 40 years, scientific attempts to model the changes we can expect to see in Xen's lifetime and later have gotten better and better. Today we have the Argo array made of long slender tubes, buoyed or lowered from ships all over the world, that drop to 2,000 meters (more than a mile deep) and profile the temperature and salinity every time they descend and rise. When they surface, they telemeter their data to satellites. Now the entire ocean can be profiled for heat and salinity in three dimensions. Add to that the data from other satellites and we have seen how the warmth and CO_2 is transferred to the ocean from moving cells in the atmosphere and how sea level is changing in response.

What we have discovered is that prior to 1980, there was very little warming below 300 meters, but then we started to see it go to 700 meters. Now we see a warmer ocean at depths of 2,000 meters. This warming is irreversible on anything approaching human timescales. Below the thermocline, ocean temperature still records past ice ages that lasted for tens or hundreds of thousands of years. To have altered that memory in a significant way in as little as 30 years is astonishing.

In 2019, an article appearing in the science journal *Nature Communications* looked at various scenarios for slow and fast changes. Relative to 2000, they found a likely average global sea rise in the twenty-first century of more than 3.3 feet (1 m) and a possible rise of more than 16.4 feet (5 m). Ohio State glaciologist Jason Box has said he believes we already have 70 feet (21.3 m) of sea-level rise baked into the system. Under even the mildest of these scenarios, sea-level rise would expose a large part of the tropics to 100-year flooding events every year from 2050. By the end of this century, it will be like that for most coastlines around the world.

In Florida, where the most far-reaching changes to cities and homes are demanded, losses from flooding may devalue homes by 15–35 percent by the time Xen is old enough to pay off a mortgage. This type of home devaluation will devastate people's savings, shred local tax revenue, and make banks unwilling to give long-term loans to do repairs or build new structures.

New York City, like other coastal cities, is already experiencing stronger storms and rising sea levels. Some residents support a $119 billion engineered floodgate in the outer New York Harbor to save people, properties, and the Statue of Liberty. Others want onshore solutions like berms, dikes, wetlands restoration, or elevated islands around lower Manhattan. Twelve of the 15 megacities are coastal.

Shanghai is another major city sinking under the weight of its development, but it is also a source of hope that good management can make a real difference. In 1995, China began requiring official permits for wells, sourcing more water from the river, and prohibiting extraction of materials for construction from within the city. That reduced Shanghai's sinking from 3.5 inches (9 cm) to approximately 0.4 inches (1 cm) per year. By pumping water back into the ground, the city was able to cause the land to rise, in some places by 4.3 inches (11 cm), which is the difference between the sea-level rise predicted for one and a half and two degrees of warming. Global temperatures and sea levels lag many years behind CO_2 increases because of ocean mixing and other factors, so we have not yet really begun to feel the inevitable heat and sea level that are

Houston, Texas
Has sunk 106 inches (3 m) and parts are sinking by 2 inches (5.8 cm) per year

London, England
The Thames Barrier, opened in 1984 to protect London from a 1-in-100-year flood, was expected to be used only rarely; it is currently used six to seven times yearly

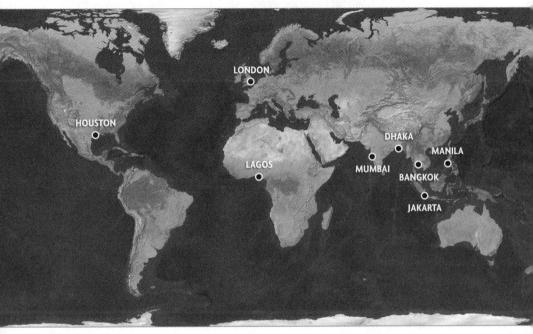

Lagos, Nigeria
Only 8 inches (20 cm) above the ocean

Mumbai, India
Underwater by 2030

Dhaka, Bangladesh
Sinking 0.55 inches (1.4 cm) per year, but sea-level rise in the Bay of Bengal is 10 times the global average

Bangkok, Thailand
Underwater by 2030

Jakarta, Indonesia
Sinking 10 inches (25.4 cm) per year, 25 times the rate of sea-level rise

Manila, Philippines
Sinking 4 inches (10 cm) per year, 10 times the rate of sea-level rise

already "baked in the cake." Earth's surface will not be fully warm until after the deep ocean waters, now about four degrees above freezing, catch up to the warmer surface waters. It will take around 1,500 years for the ocean to mix and turn over enough to reflect the temperature of surface water today, but by then, surface water may be far hotter still. Most projections of temperature and sea level only look at about 10 percent of the effect because they limit predictions to only a century or two.

Perhaps the best way to project what kinds of changes we can eventually expect from 400 parts per million in the atmosphere, or two degrees of air temperature warming, would be to look to the distant past when Earth last experienced those conditions. Biogeophysicist Thomas J. F. Goreau notes that "the last time that global temperatures were 1°C to 2°C [1.8°F to 3.6°F] above today's level, sea levels were about 8 meters (26 feet) higher, crocodiles and hippopotamuses lived in swamps where London now stands, and CO_2 levels were around 270 ppm, approximately 40 percent lower than today." At a CO_2 concentration of 400 ppm, once the climate system had fully responded, air temperatures would be about 17°C (30.6°F) warmer, and sea levels would be 23 meters (75 feet) above today's levels, Goreau says.

We are committed to such changes, even if there is no further CO_2 increase, starting right now, because of the excess already in the atmosphere, unless that is rapidly reduced. No amount of emissions reduction, such as by switching to solar-powered automobiles, having fewer pets, or eating less grain-fed meat, could reduce excess atmospheric CO_2. Only increased natural carbon sinks can draw down the dangerous excess in time to avert extreme long-term changes that will last for hundreds of thousands to millions of years.

Perhaps the most significant cost of adaptation to climate change will be the cost of protecting low-lying islands and coasts from being flooded by global sea-level rise. Most sea-level change does not occur in gradual steps, where each year, beachfront homes are a fraction of an inch closer to the water. Change comes in spurts when storms combine with conditions like full-moon tides or low-pressure cells to push water around and effect permanent changes to coasts. Global warming of ocean water will cause higher storm surges and wind-made waves and will feed the strength of hurricanes and typhoons.

Half of South Florida is less than 3 feet (0.91 m) above sea level. Florida is built upon porous limestone that saltwater can penetrate. Along the coast, some of the world's most expensive homes are on or near the beach, as well as 75 percent of the 5.5 million people in South Florida. With more than $416 billion in city assets at risk, Miami is ground zero.

Miami, for thousands of years, was uninhabited, even by Native peoples, because it had been a mosquito, snake, and alligator swamp. In the late 1800s, city planners arrived, canals were dug, swamps were drained, and in the 1920s, a great city emerged—much like New Orleans and Washington, DC—built upon coastal wetlands.

Because the young city was at a strategic maritime crossroads with Cuba, the Caribbean, and Latin America, it prospered and grew. Eager newcomers chopped down the mangroves and started building on the barrier islands to its Atlantic side and pushed farther into the swampy lowlands to its Gulf side.

Miami Beach is now among the most flood-prone areas in the world. Any full moon can push water backward through the old storm drains and flood the streets and sidewalks. On a very high tide, Miami Beach has surfable waves on its streets. As the ocean rises, this will get worse each year, until the city disappears entirely and becomes an underwater attraction for scuba divers. Along with the city will go Everglades National Park, a world biodiversity treasure.

The people who now live in South Florida, along with farms and other businesses, consume some three billion gallons of drinking water every day, pumped from a freshwater aquifer. As the aquifer depletes its million-year-old supply, saltwater has been creeping in, forcing the city to install expensive pumps to hold back the salt. Over the long term, saltwater cannot be kept from ruining the aquifer. And, no matter how much concrete is poured to raise the city above the waves, it cannot survive without water.

There is another concern that could be as big for Miami as its water. Just outside the city, at a stretch of beach called Turkey Point, there are two nuclear reactors that were built in the 1970s. To protect them from hurricanes, the reactor buildings are elevated 20 feet (6 m) above sea level, several feet above what was thought to be the maximum storm surge 50 years ago. Hurricane Andrew, a Category 5 hurricane that passed 10 miles north of Turkey Point in 1992, had a storm surge 3 feet (0.91 m) high, but not far away, on the back edge of the storm, Andrew's flooding was 17 feet (5 m). Three feet of ocean flooding cut off Turkey Point from the mainland so it could only be reached by boat or helicopter. Turkey Point's emergency diesel generators are located at 15 feet (4.6 m).

It was the failure of emergency diesel generators to keep cooling waters circulating (because they could not operate underwater) that caused the catastrophic explosion and meltdown of three reactors at Fukushima in 2011. The higher the seas go, the deeper emergency generators could be submerged. Category 4 Hurricane Katrina hit New Orleans in 2005 with a storm surge of 28 feet (8.5 m). Despite all of these concerns, Turkey Point's owner, Florida Power & Light Company, is proposing to spend $18 billion to build two more reactors at the same location that would operate through at least 2085.

12

Marine Heat Waves

A marine heat wave is defined as when daily sea surface temperatures are warmer than 99 percent of previous observations for the same time of the year over the baseline period from 1982 to 2016. At the end of 2019 and the start of 2020, the waters off Western Australia met that threshold, with temperatures 2°C (3.6°F) warmer than ever before observed for the Christmas to New Year period. As hot as that water was, the temperatures were not as high as the 2011 heat wave, which occurred later in summer. An intense marine heat wave off the east coast of Tasmania persisted for 251 days—from spring 2015 to autumn 2016.

It should be no surprise, even for a continent with 400 years of recorded wildfires, that Australia's catastrophic 2019–2020 fire season corresponded with a massive offshore heat wave that fueled drought and then-unparalleled bushfire ferocity on land. At least one billion wild animals are estimated to have died (not including frogs and insects), and some may be facing extinction.

What was unseen—hidden in the dark and silent depths—was the marine death toll. In 2013, a large patch of hot water in the north-east Pacific Ocean that scientists dubbed "the Blob" began forming.

HURRICANES

Few things in nature have the power of an atomic bomb, but tropical cyclones do. These storms, which are called hurricanes in the Atlantic and typhoons in the Pacific, begin when a column of hot air rises off of the ocean and, because of wind conditions at higher altitudes, begins to rotate in a wide spinning motion. The size of these hot air cells can be enormous. Photographs from space peering down at Hurricane Wilma in 2005 show a cloud spiral half as large as the Caribbean Sea. Hurricane Harvey in 2017 blanketed the entire Gulf of Mexico.

In the two years of the Northeast Pacific heat wave, a persistent high-pressure ridge prevented normal winter storms from reaching California, bringing a devastating drought from 2013 to 2015 but being too far north to generate a hurricane. Similarly, heat waves in the Tasman Sea in 2015–16 and coastal Peru in 2017 caused heavy rainfall and flooding but were too far from the equator for hurricanes.

As the ocean warms and heating also makes the atmosphere more turbulent, we can expect to see the conditions needed to trigger hurricanes expand from latitudes close to the equator to a much wider area. Hurricane season, which corresponds to summer warming of the ocean, will also get longer. Marine heat waves will make the surface of the ocean hotter, making conditions ripe for massive hurricanes.

It grew more intense in 2015 under the surface warming conditions known as El Niño. About 62,000 dead or dying common murres, the dominant fish-eating seabird of the North Pacific, washed ashore between summer 2015 and spring 2016 on beaches from California to Alaska. Because of the unseen carcasses that never washed ashore, researchers have recently put the murres' death toll closer to one million. Because cod, pollock, halibut, and hake eat many of the same foods as murres, the blob also brought mass mortality to them. In the case of cod, the starvation loss is estimated at 100 million. Some 100 humpback whales that subsist on krill and herring also starved.

At first, scientists thought this was a fluke event, something that might occur once in centuries. University climatologist Nick Bond told reporters that "the original blob was so unusual, and stood above the usual kind of variations in the climate and ocean temperatures, that we thought 'wow, this is going to be something we won't see for quite a while.'" Then, in December 2019, something scientists are now calling "Blob 2.0" was reported once more to be forming in the North

The costliest hurricane season on record, 2017, included storms Harvey, Irma, and Maria, which collectively caused $265 billion in damage. They were just an indicator of what is still to come. Climate change leading up to 2017 had caused a warmer Gulf of Mexico that increased the rainfall intensity by 8–19 percent. Hurricane Harvey rained down super-torrents on Texas and produced a storm surge of more than 6.6 feet (2 m) in coastal areas. Those floods released 4.6 million pounds of toxic contaminants from petrochemical plants and refineries in Houston, spreading deadly poisons far and wide.

Hurricane Irma's 160-knot winds wiped out housing, schools, fisheries, and farms in Barbuda, Antigua, St. Martin, and the British Virgin Islands. Hurricane Maria, which trailed a few days behind with winds of 150 knots, made landfall on Dominica, Puerto Rico, and Turks and Caicos, destroying almost all power lines, buildings, and 80 percent of crops in Puerto Rico. It damaged nearly 100 percent of Puerto Rico's economy. It completely evacuated Barbuda. The full number of casualties from the storm may never be known.

Climate change will increase the frequency and intensity of such storms in the future and magnify impacts on those who live near the sea.

Pacific. It ranges from a degree or two warmer at its extremities to as much as four or five degrees warmer at its core.

The ocean absorbs most of the heating effect of greenhouse gas pollution—90 percent of the excess heat in the climate system—and it is having an effect. Since the ocean's temperature was first taken with modern instruments half a century ago, the past five years are the top five warmest, and the past ten years are also the top ten years on record.

Nonetheless, since 1993, the *rate* of ocean warming has more than doubled. Marine heat waves have doubled in frequency since 1982 and are increasing in intensity every year. They have increased in both frequency and intensity over the past 35 years, and climate change is projected to increase heat waves in the future and all their cascading impacts. During Xen's lifetime, they will soak up seven times more heat than they did in the half century before Xen was born. Climate models project increases in the frequency of marine heat waves by 20 to 50 times. Their intensity is projected to increase about tenfold. The most substantial increases in frequency are expected in the Arctic and the tropical oceans.

Scientists measured marine heat waves in the Mediterranean Sea in 2003; Western Australia, 2011; Northwest Atlantic, 2012; Northeast Pacific, 2013–15; Sea of Japan, 2016; equatorial Pacific, 2014–17; Tasman Sea, 2015–16; coastal Peru, 2017; and Southwest Atlantic, 2017. Regional high-resolution climate simulations suggest that the Mediterranean Sea will experience at least one long-lasting heat wave every year by the end of this century.

The heat stress during the three-year equatorial Pacific event was sufficient to cause bleaching of 75 percent of all the coral reefs in the world and extermination of 30 percent of them—more than any previous bleaching event. Seagrass meadows and kelp forests showed similar damage.

The Northeast Pacific 2013–2015 "blob" caused a coast-wide bloom of the toxigenic diatom *Pseudo-nitzschia* that resulted in the largest influx of domoic acid ever recorded along the west coast of North America, threatening whales, dolphins, porpoises, seals, and sea lions. Elevated toxins detected in commercially harvested fish, shrimp, and crabs resulted in prolonged closure of razor clam and crab fisheries. The coastal Peruvian heat wave in 2017 caused anchoveta to have decreased fat content and take to early spawning to survive.

The 2019 Christmas heat wave in Australia killed crabs, shellfish and mollusks (such as oysters and abalone), and krill at the mouths of rivers. These were deaths that were easily observed from the land. The effects on coral reefs and seagrass meadows were not as visible but were potentially more deadly.

What starts in the sea doesn't stay in the sea. As hot water evaporates, it moistens the air above the sea surface. In 2020, the air above the oceans was 5–15 percent moister, on average, than it was in 1970. Storms grab that moisture to make more-powerful hurricanes and typhoons and give them more-extreme rains and flooding. Hurricane Harvey in August 2017 was produced by a Gulf of Mexico heat wave that it encountered after crossing the Yucatan peninsula. It dropped more than 5 feet (1.5 m) of rain on Nederland, Texas, half of that in a single 24-hour period. An estimated 300,000 structures and 500,000 vehicles were damaged or destroyed in the Houston area.

Ocean heat content in 2019 was the warmest on record by a sizable margin. Between 2018 and 2019, the oceans absorbed an amount of heat around four times larger than all the energy used by humans in the world, the equivalent of one atomic bomb per second. According

THE CORAL TRIANGLE

The Coral Triangle covers four million square miles of ocean and coastal waters in Southeast Asia and the Pacific, in the area surrounding Indonesia, Malaysia, Papua New Guinea, the Philippines, Timor-Leste, and the Solomon Islands. It is the center of the highest coastal marine biodiversity in the world. But because of these riches, human activities are also expanding. There is coastal development to accommodate a booming tourism sector and overfishing to supply the restaurants that feed these tourists. The Coral Triangle has already lost nearly half of its coral reefs and mangroves over the past 40 years. Sea surface temperatures grew by 0.1°C (32.2°F) per decade after scientists began taking measurements in 1960, but by 2006 the rate of warming had increased to 0.2°C (32.4°F). If that kind of increase, or accelerated heating, continues, the heat waves will only get worse and come more often.

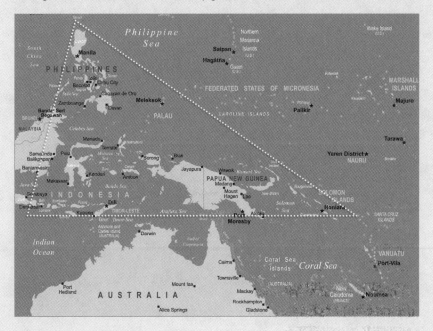

to an analysis by the Grantham Research Institute, if the same amount of heat had gone into the lower six miles of the atmosphere that went into the top mile of the ocean between 1955 and 2010, the Earth would have seen a warming of 36°C (64.8°F).

That is to say, and not to put too fine a point on it, without the ocean buffering that warmth, you and I would no longer be here.

PART FIVE

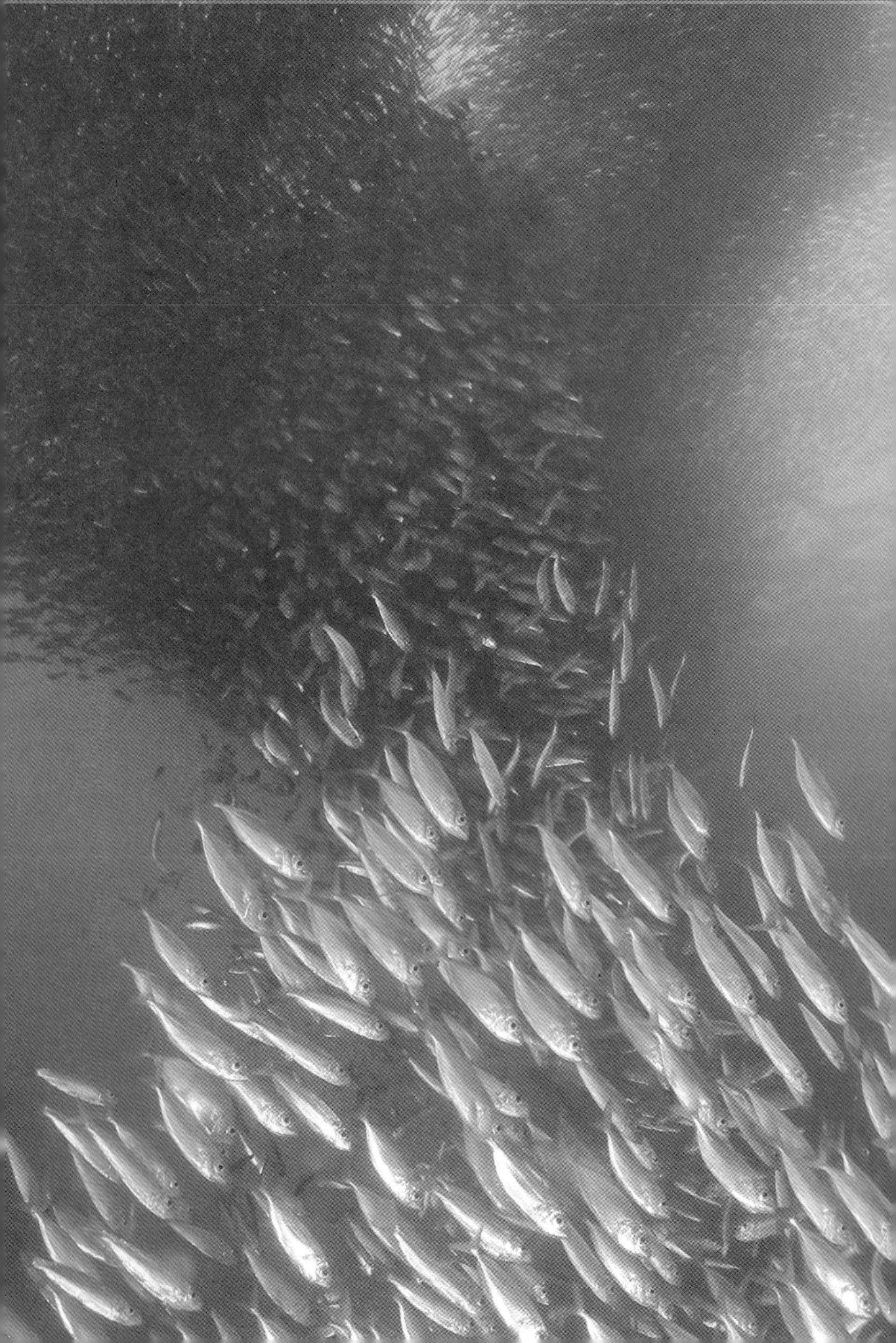

CHAPTER

13

Geoengineering

When he saw what the loss of plankton was doing to the annual salmon run, Russell George had an idea. His concept was that iron spread upon the waters could fertilize plankton blooms. The plankton would feed fish, making more and fatter salmon, and at the same time draw carbon dioxide from the atmosphere as the plankton thrived, died, and then sank to the ocean floor.

Scientific review bodies, such as the UK's Royal Society and the National Oceanic and Atmospheric Administration, were quick to throw cold water on this idea. Most iron uptake is only temporary, and any plan to fertilize the ocean would be expensive and short-lived. Worse, these panels of scientists worried, there could be unintended consequences, and some of those could be far removed in space and time. With no regulatory framework in place, let alone a scientific protocol, experts recommended against George's experiment.

George reasoned that there is no regulatory framework in place, let alone a scientific protocol, so nothing was stopping him. In July 2012, 200,000 pounds (90.7 MT) of iron sulfate were dumped into the northeast Pacific Ocean, paid for by salmon fishermen. As it turned out, there was a regulatory framework in place. The dumping violated the United Nations Convention on Biological Diversity and the London Convention on the Prevention of Marine Pollution by Dumping of Wastes and Other Matter. The enforcement branch of Environment and Climate Change Canada executed warrants on Russell George, and the experiment ended.

In part because of Russell George, the UN Convention on the Law of the Sea now refers explicitly to geoengineering following an amendment adopted in 2013 via resolution LP.4(8) adding a new Article 6bis:

> Contracting Parties shall not allow the placement of matter into the sea from vessels, aircraft, platforms or other man-made structures at sea for marine geoengineering activities . . . unless . . . authorized under a permit.

Then, in 2013, the west coast of North America experienced its largest salmon return and, subsequently, its most abundant commercial salmon harvest in history, jumping from 50 million to 226 million fish. Commercial fisheries opened in areas that had not seen a commercial opening since the 1960s. In 2014, the Fraser River experienced its second-largest sockeye salmon return in history, while the Columbia River recorded its most extensive salmon run of all time.

We don't know whether those salmon were a result of Russell George's experiment or not, but there is another question that should be asked: How nutritious were those salmon?

Researchers at Arizona State University who gave plankton more light and minerals to speed growth discovered that rather than boosting food supply—both for humans and the marine food web—chemical fertilization may hurt it. Increased food (such as iron) made plankton grow faster, but it ended up containing fewer of the nutrients the plankton needed to thrive. By speeding up their growth, the researchers had essentially turned the plankton into junk food. They had plenty to eat, but their food was less nutritious, so they were starving. The same effect moved up the food chain. Turtles and fish ate the plankton, and dolphins and whales ate the fish, but none was getting much nutrition from what it consumed. Everything along the chain was undernourished.

Since at least the 1980s, there have been ideas tossed around about ways that humans might engineer their way out of the climate emergency, including ocean geoengineering, but many arrive at the same dead end: unintended consequences. For instance, large-scale fertilization of the ocean could release climate-relevant gases like methane, cause eutrophication due to deep water nutrients coming to the surface, create more low-oxygen zones, add to ocean acidification, and increase toxic phytoplankton populations. By altering ecosystem structure, these changes would likely affect fisheries.

Other geoengineering ideas regarded as too risky are *assisted evolution* via genetic alterations to species of marine life, *enhancing weathering* and *alkalinization* by adding CO_2-absorbing materials to the ocean, and the sunlight reflection techniques of marine *cloud brightening* and *surface albedo enhancement*.

One proposal being bounced around lately is fizzy water. On ocean surfaces, bubbles have similar reflective properties to the droplets formed in clouds, so the idea is to enhance the Earth's albedo by generating microbubbles to brighten the ocean and have a localized cooling effect. This could be accomplished simply by reconfiguring ship propellers to enhance the sparkle of their wakes. The alteration might have the added benefit of increasing oxygen content in surface waters, or it could have the negative effect of causing CO_2 outgassing to the atmosphere.

While many of these schemes have a theoretical potential to slow climate change, only George's plankton fertilization experiment has been tested in the field, with unproven success. Much more attention needs to be given to careful study, international governance, and public acceptability before these proposals get more open ocean trials.

In the early days of biotechnology, ethical concerns about our right to manipulate complex organisms were not listened to by biotech companies and government regulators. Today, in aquaculture and open-sea fish farming, it is common to introduce genetically modified fish that can better resist disease, gain weight more rapidly, or adapt to restricted diets. Although the farms engineer barriers to prevent accidental release and mingling with wild populations, it is not unheard of for some of these GM fish to escape into the wild. By not thinking carefully about the consequences, we open possibilities we may no longer be able to control.

One reason fish farmers want to short-cut the breeding process (a process evolved over a billion years, developing safeguards along the way) is because feed is expensive. Bycatch fish meal and oil from wild sources like anchoveta, herring, or sardines is costly, and prices have risen as demand has grown. About half of the fish meal and nearly all the fish oil produced globally is now consumed by aquaculture, with the rest going to pigs, hens, and cows. A natural, non-GMO, farmed salmon needs 5 pounds (2 kg) of pressed anchovy pellets to produce a pound of fine red flesh.

The large-scale removal of small pelagic feed-fish (e.g., sprats, sardines, herrings) impacts both their prey and predators. Feed-fish are primarily caught using seine nets that may have high levels of bycatch. Small pelagics are nutritionally rich and of crucial importance in the diets of wild fish populations. Overfishing of feed-fish species in coastal areas can result in reduced genetic diversity and loss of adaptive capacity to bigger threats like climate change.

Growers tried feeding salmon canola and soy oils, but the health and flavor of the fish declined. Most recently, the agrochem giant BASF has developed GMO canola plants (canola itself is laboratory-bred rapeseed—the name stands for "Canadian oil") to produce marine-like oils high in healthy omega-3 fatty acids. Other companies are breeding varieties of fish that can better tolerate a plant-based diet.

Close confinement is another problem that fish farms have to confront because it is not the natural condition of wild animals. In tight conditions, they grow weaker and more prone to infectious diseases. Antibiotics are routinely administered—in Norway, every farmed fish is vaccinated—but there is a danger to this approach because eventually viruses and bacteria breed out vulnerabilities to introduced antibodies and become resistant. In the 1990s, a strain of cholera that had picked up antibiotic resistance from drugs used in Ecuadorian shrimp farms made its way around South America. To counter this difficulty, breeding programs are skipping necessary safeguards and introducing newly engineered, disease-resistant fish into the farms and, occasionally, the ocean.

In 2020, US consumers got their first taste of AquAdvantage salmon, a genetically engineered Atlantic salmon developed by AquaBounty Technologies in 1989. The typical growth-hormone-regulating gene in the Atlantic salmon was replaced with the growth-hormone-regulating gene from Pacific Chinook salmon, along with another sequence from ocean pout, an eel-like Atlantic fish with blood proteins that work as antifreeze. Besides taking extreme cold, this "Frankenfish" (as Alaska senator Lisa Murkowski called it) grows faster than conventional salmon and therefore gets to market weight in less time, with less feeding. Aqua-Bounty is growing its engineered fish in land-based freshwater tanks in a US-inspected facility in the mountains of Panama but plans to open a large farm in Indiana for the US market. An added protection engineered in is that the females are all sterile and cannot breed in the wild. The fertile broodstock females are raised in a closed facility on Prince Edward Island in Canada.

Mixing genetically modified populations with wild populations is a real concern because the newcomers may outcompete native populations. Parasites and pathogens have also crossed over to native populations, most notably in areas with both wild salmon fisheries and ocean farm production. The overuse and inappropriate application of antibiotics contribute to the global spread of antimicrobial resistance, while excess pesticides, nitrogen, and other byproducts from enclosures pollute surrounding waters.

At the beginning of this food revolution, we never paused to consider whether these genetically modified and farmed fish might have an opinion about the matter. Would they prefer living in the wild and migrating in schools thousands of miles every year, or spending their entire lives caged in close proximity to hundreds of their brethren? Science critic Jeff VanderMeer says:

> Whether we were affixing an ear to a mouse or growing miniature chimp brains in ostrich bodies and then shoving those brains into dinosaurs, we did not want to think overmuch about the individual animals we were experimenting on—or even the more robust creations that came after those first and second generations. By not thinking carefully about the consequences, we abandoned any moral high ground and created a situation in which we may soon be unsure that we control our own minds—as individuals or in aggregate as human beings on this fragile globe.

Film director Nora Bateson says:

> Our way of thinking about nature is deeply rooted in a particular kind of patterning that is suited to receive only some of the information around us. And so, when we start to come to conclusions and find solutions and link things together, there is a need, first and foremost, to recognize, "I am only capable of seeing one very small portion of what is happening around me."
>
> We are only studying nature though the ways in which the human species can perceive nature after the last thousand years of the development of Western Civilization.

Over the centuries, various geoengineering schemes have been proposed to either heat or cool the North Pole. Today, most would attempt to slow the melting of ice at the North Pole by spraying aerosols to reflect sunlight or thickening ice by spraying seawater onto existing surfaces. At the beginning of the twentieth century, American electrical wizard Charles Proteus Steinmetz proposed to widen the Bering Strait to 200 miles by removing St. Lawrence Island and parts of Seward and Chukchi Peninsulas to let the Japan Current warm the Arctic Ocean.

After the Second World War, the Soviet Union proposed a joint project with the US to warm the Arctic Ocean and melt some of the ice caps to permit navigation. The 1956 design by Petr Borisov called for erecting a dam across the narrowest part of the strait between Alaska and Russia, a distance of about 60 miles, with depths up to 160 feet (50 meters) and extreme currents. By pumping low-salinity cold surface water across the dam to the Pacific, warmer and higher salinity seawater from the Atlantic would flow into the Arctic. However, citing national security concerns, the CIA and FBI opposed the Soviet plan, saying that even if the project were feasible, it would compromise the North American Aerospace Defense Command. Soviet scientist D. A. Drogaytsev also opposed the idea, arguing that the sea north of the dam and north-flowing rivers in Siberia would become unnavigable year-round and that the Gobi and other deserts would extend to the northern Siberia coastline.

Most recently, it has been suggested that Arctic sea ice could be preserved by influencing salinity and temperature of the Arctic Ocean by changing the ratio of Pacific and Yukon River waters entering through the Bering Strait, which, paradoxically, is precisely what was proposed in 1956 to get the opposite effect—to warm the Arctic for navigation.

These projects would be almost unimaginably expensive, would have to be continued indefinitely, and would have no guarantee of success. Worse, as D. A. Drogaytsev pointed out in 1956, either failure or success could bring far worse consequences than just leaving nature alone.

14

Mining the Seabed

The debate over whether deep-sea mining has a place in an environmentally and socially sustainable "blue" economy is heating up. Proponents argue that we will need resources from the ocean to transition to a low-carbon economy. Metal demand for electric-vehicle batteries, for instance, will have to increase more than tenfold by 2050 to meet the requirements of the Paris Agreement to limit global warming. Opponents fear mining the seafloor will devastate the last untouched wilderness on the planet.

Potato-sized polymetallic nodules, which contain nickel, cobalt, copper, and manganese, lie on the seabed at depths of 2.5–3.7 miles (4–6 km) in an area of the Pacific called the Clarion-Clipperton Zone. These ores are in demand for batteries and wiring in electric vehicles, windmills, and solar panels. A dozen countries, including China, India, Japan, Russia, and the UK, have been granted exploration contracts regulated by the International Seabed Authority (ISA).

These countries plan to send deep-diving drone miners the size of a combine harvester trawling the seabeds to remove the top layer of sediment and pump it through a pipeline to a ship, which then separates the nodules and discharges the sediment into the ocean.

One concern is that the sediment plume of processed materials and wastewater could carry for great distances, suffocating marine life and impacting ecosystems far beyond the mining area. Near-shore dredging accelerates beach erosion by reducing the sediment supply to the coast. Noise, light, and chemical pollution from mining vehicles also affect marine organisms.

Another concern is that the mineral-rich discharges could feed algal blooms at the surface. Still another is that the quantity and diversity of biological species in the deep sea are far higher than previously thought, and, with the direct removal of seabed fauna and flora, entire ecosystems could be gone before many species are even named. Potential impacts on fragile hydrothermal vents, for instance, include habitat destruction, rare species extinction, and modification of fluid flux regimes.

Michael Lodge, secretary-general of the ISA, which should be the agency protecting the ocean, says: "If you said that no industry can start until we know what is going to happen from that industry, then that's an entirely circular argument that would prevent any industry in the history of humanity from starting.

"We have a good idea of what the impacts will be," Mr. Lodge says. "They are by no means as catastrophic as environmental groups would have us believe; they are predictable and manageable." His regulatory body argues that biodiversity losses from surface mining are likely to be much worse, given the greater abundance of wildlife in many mountainous areas.

Exploratory licenses have been granted for more than 1.3 million square kilometers (500,000 square miles) of the seabed in areas beyond national jurisdiction. While the Law of the Sea does not presently extend to this area, international regulations are expected to be created over the coming years. The ISA will be tasked with policing. In the meantime, ISA has issued mining leases to Germany (2015–2030) and China (2011–2026) to mine seabeds rich in gold, silver, copper, zinc, and lead.

With global recycling rates for electronic waste at only around 20 percent, a large amount of valuable metals that could go into electric cars and wind generators is being wasted. The Deep Sea Conservation Coalition says we should be talking about reusing and recycling what we've already got rather than opening up a whole new frontier of environmental degradation to feed a throwaway economy. Given the potential losses just to genetic wealth, some restraint would seem to be in order.

The industry agrees that recycling should be maximized, but says that will not supply the substantial additional volume of metal needed to manufacture a billion new electric vehicles. "You can't recycle what you don't have," a spokesman for one manufacturer says. "What we, first of all, need to do is to have a massive injection of new battery materials put into the system."

Jean-Baptiste Jouffray, working at the Stockholm Resilience Centre, has described the attack on the oceans since 2000 as the "blue acceleration." In a science journal published in January 2020, he observed that as demand for resources continues to grow and land-based non-renewable resources decline, expectations for the ocean as an engine of human development are increasing. The extent, intensity, and diversity of today's aspirations are unprecedented. One example Jouffray gives is the scaly-foot snail (*Chrysomallon squamiferum*), first discovered in 1999, named in 2015, and by July 2019 already on the Red List of Threatened Species:

> Found more than 2,400 m beneath the ocean's surface, on just three deep-sea hydrothermal vent systems that collectively cover an area of twenty square meters, the scaly-foot snail's future was deemed threatened when two of these vent systems fell within exploratory mining leases granted by the International Seabed Authority. . . . Because of its unique, tri-layered natural armor, the scaly-foot snail has been the focus of biomimicry research funded by the US Department of Defense, and 118 sequences from its genome have been deposited in GenBank, an open-access database of nucleotide sequences that serves as a reference point for the biotechnology industry.
>
> The claims on the scaly-foot snail, therefore, extend from its surrounding habitat to its physical form and genetic information through to its own existence. This example illustrates several dynamics of the blue acceleration. All the metals found in seafloor massive sulfides can be mined on land, but demand for use in high-end electronics and a decline in the ore quality of land-based sources have caused the commodity values to rapidly increase since 2000 (e.g., gold, +454%; silver, +317%; copper, +360%; zinc, +259%; lead, +493%), making seabed mining a viable commercial prospect. Contrary to the precautionary principle, exploitation is proceeding ahead of exploration,

with mining licenses granted prior to a consensus on how to mitigate environmental impacts of mining, and despite the three hydrothermal vent systems (Solitaire, Kairei, and Longqi fields) not yet having been studied in detail. Finally, because of the placement of claims and the complexity of territorial boundaries, the survival of the scaly-foot snail moved within five years from a responsibility of the global community to the responsibility of three countries: China, Germany, and Mauritius.

Which would we rather have—more electric cars or more scaly-foot snails? What do we do when reversing climate change conflicts with preserving biodiversity? When we speak of sustainability, what is it we are trying to sustain? Our ability to supply fish oils to cats and cattle? A seafood-consuming human population of 8 billion and counting? Or might we rather have, at the end of it all, the web of life that provides the air we breathe and balances the temperature of our planet to within the range we require to survive?

BIOPROSPECTING

Only 5 percent of the seafloor has been mapped to the level of detail of the Moon and Mars. The deep sea is one of the least explored and most extreme environments on Earth. Pressure increases by one atmosphere for every 10 meters deeper you go—to greater than 1,000 atmospheres in the deepest parts of the trenches. Temperature hovers just above freezing on the abyssal plain and near boiling beside thermal vents. It was only with the advent of remotely operated vehicles (ROV) and submersibles that the deep ocean unveiled its mysteries, including stunning microbial biodiversity. Like octopuses and jellyfish, microorganisms inhabiting these harsh environments developed unique strategies to survive, especially to the high pressure. Their adaptation to biochemical and physiological processes is mirrored in modifications to gene expressions that produce novel natural products of use to society. Given the size of the ocean, it is possible the biodiversity is higher there than on land, and because of the greater extremes, epigenetic adaptations could be even more numerous.

Satellite imagery, DNA mapping, and subsea sampling instruments are opening up new frontiers. Genetic modification technologies have dropped in cost by four orders of magnitude over just the past 10 years. The amount of data flowing into the World Ocean

Database, the most extensive publicly available, uniform format, quality-controlled, global ocean profile dataset, is growing exponentially, with millions of reference points added every year.

In 2008, a Yunnan University scientist, Xin-Peng Tian, and his coworkers peered into a soft black mud pulled up from 12,680 feet (3,865 m) below the northern South China Sea and identified a new genus of spore-forming actinomycete, or bacterium, *Marinactinospora gen. nov.* Polypeptides synthesized from one specimen, *Marinactinospora thermotolerans*, exhibited vigorous antibacterial activity against *Micrococcus luteus*, a common source of hospital infection; *Staphylococcus aureus*, a cause of respiratory and skin infections and food-poisoning, lately showing resistance to antibacterial treatment; *Bacillus subtilis*, a common gut bacterium; and *B. thuringiensis*, a bacterium taken from caterpillars and moths and commonly used to make food crops more insect-resistant. The *M. thermotolerans* strain itself was resistant to cold, amoxicillin, ampicillin, lincomycin, penicillin G, and streptomycin. In other words, it had huge potential commercial value to medicine.

Since the first compounds were extracted from the Caribbean marine sponge *Cryptotethya crypta* in the early 1950s, biotech firms have been swarming to gather patents on genes, the natural compounds they code for, and the organisms that reproduce them. Today, over 34,000 natural products derive from species found in the ocean. Marine organisms are of particular interest for pharmaceuticals, nutraceuticals, cosmeceuticals, and chemicals. As of August 2018, 13,171 genetic sequences from 865 marine species had been issued international protection under the Patent Cooperation Treaty.

The Nagoya Protocol on Access to Genetic Resources and the Fair and Equitable Sharing of Benefits Arising from their Utilization to the Convention on Biological Diversity (signed in 2010; entered into force in 2014) is the key reference point for governing access to genetic resources and subsequent sharing of benefits from their use. Provider countries (where genetic resources are located) and user countries (those seeking access to these genetic resources) must arrive at mutually agreed terms (Article 18) based on prior informed consent (Article 6). In areas beyond national jurisdiction, no limitations are placed on the collection of marine genetic resources or their subsequent use. This governance gap is a subject of the ongoing negotiations for amendments to the Law of the Sea. The UN General Assembly has called for conservation and sustainable use of marine biological diversity.

PART SIX

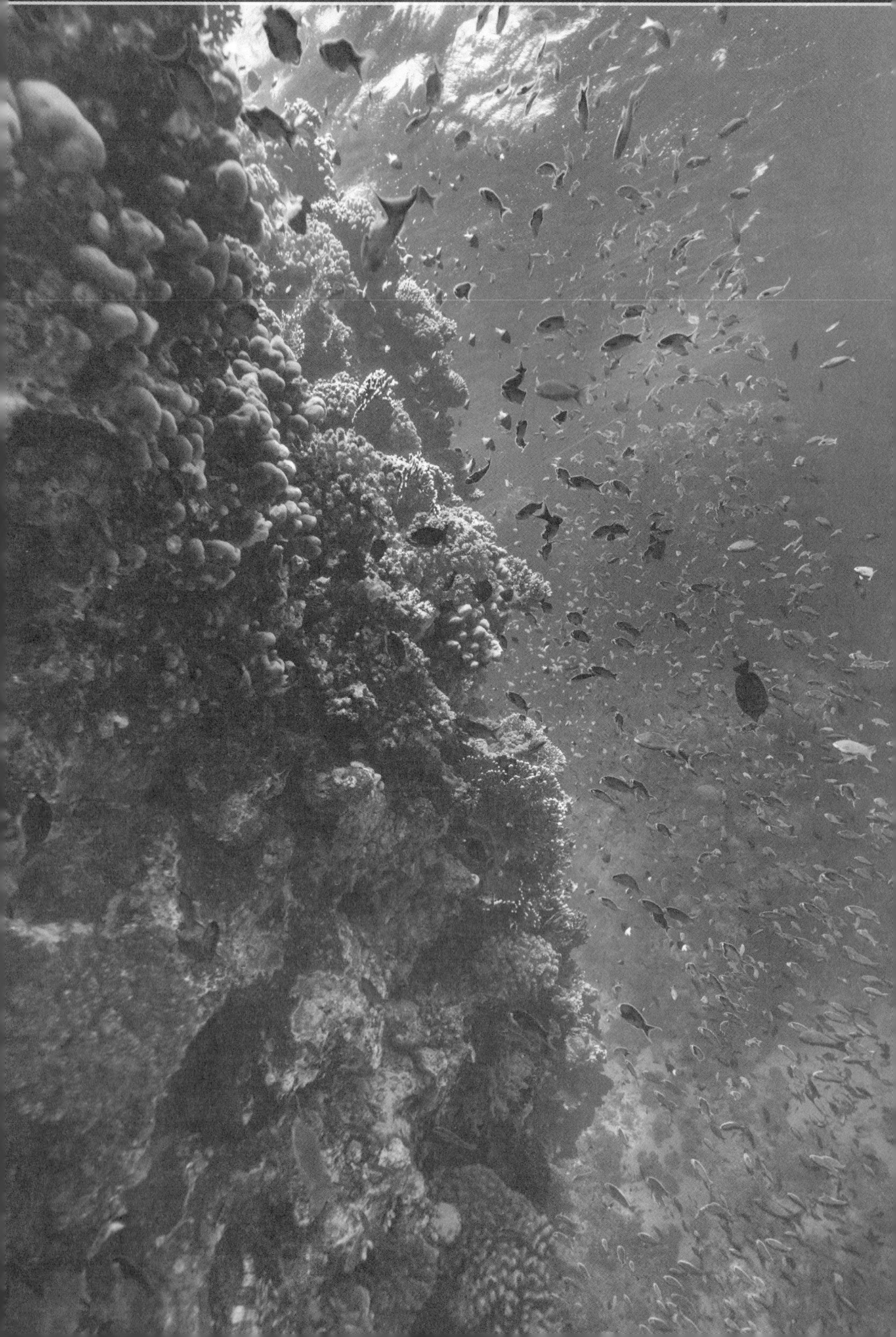

CHAPTER

15

Ridge to Reef

Biogeophysicist Thomas J. F. Goreau cuts a striking figure, whether he is coming out of the water in a wetsuit and scuba or dressed in a formal waistcoat, strolling through a palace, and chatting with a member of a royal family. His white curly hair and beard frame a tanned face that is always smiling. His compact frame is muscular, befitting the swimmer he has been since diving the Jamaican reefs as a child.

He was just 19 when his father, Thomas Fritz Goreau, died at age 45, likely from radiation poisoning received during years of study of the corals of Bikini Atoll for the Atomic Energy Commission. Perhaps he was the diver who speared that radioactive fish that persuaded Admiral Blandy to end Operation Crossroads.

Tom's grandfather, who had also dived those reefs, was one of the world's earliest marine biologists. He left Germany just before Adolf Hitler's appointment as chancellor in 1933, when Tom's father was eight. Tom's father took doctorates in ecology and medicine and went on to pioneer the use of scuba gear as a marine research tool and build his own rebreather gear to explore the deep sea. In 1949 he met and married Nora Arango and the two became the world's leading researchers in the anatomy, physiology, biochemistry, and histology of coral reefs. The younger Tom, born in 1950, followed his parents' footsteps, founding the Global Coral Reef Alliance in Jamaica and becoming a world expert in marine biology. He was also Senior Scientific Affairs Officer at the UN Commission for Science and Technology for Development.

One of Tom Goreau's passions is rebuilding coral reefs, even as all around the world they are dying from heat and acid seas. Besides being biological centers, Tom says coral reefs provide perfect natural shore protection, dissipating around 97 percent of wave energy before it reaches land. As the waves pound on the reefs, coral sand is produced and transported to build beaches behind the reef rapidly. Healthy reefs are sand factories, generating vast amounts of new sand—the skeletal remains of calcareous green and red algae. Every grain of white sand on a tropical beach was once a living coral reef organism. The reef itself is self-repairing. Because of the mass mortality of corals around the world, tropical beaches that were growing until recently have begun eroding. Barrier islands are washing away, leaving coastlines more vulnerable to sea-level rise and storms.

Goreau has two secret weapons that make miracles possible—biochar and biorock. Biochar is carbon that remains in a hard, mineral form when any plant or other carbonaceous material is raised to a high temperature in the absence of oxygen. This hard carbon rock, ground into a powder, works like a coral reef in soils, holding air and water, conducting an electrical charge that stores food for plants, and becoming a popular habitat for soil microbes and fungi. Placed along watercourses where runoff enters rivers or the ocean, biochar filters the water so that dead zones and algal blooms are less likely.

Biorock comes from a discovery by Goreau's late research partner, Wolf Hilbertz, who nearly 50 years ago was studying how seashells and reefs grow by passing electric currents through saltwater. In 1974, he discovered that as the saltwater naturally electrolyzes, calcium carbonate (aragonite) combines with magnesium, chloride, and hydroxyl ions to slowly accrete around some core structure, eventually building a material skeleton stronger than most concretes.

Since Hilbertz's death in 2007, Goreau has continued using biorock to restore coral reefs, while advocating for biochar to restore the forested ridges that feed them. Applying a low-voltage electric current (safe for swimmers and marine life) to a submerged, conductive structure like iron rods or mesh fabric, Goreau causes dissolved minerals in seawater to precipitate and form a composite of brucite hydromagnesite and limestone, similar to the composition of natural coral reefs and sandy beaches. Live corals can then be transplanted onto this structure or will find and colonize it from the wild.

Tom Goreau's biorock coral reef projects have been installed in over 20 countries, in the Caribbean, Indian Ocean, Pacific, and Southeast Asia. Goreau is about to start new biorock projects to increase mangrove carbon storage (as peat) in Borneo. Biorock is cost-effective, requiring only some metal bars and electricity. Projects typically use renewable solar power, wind power, tidal power, or wave power to maintain the current. Biorock accelerates growth on coral reefs by as much as five-fold and increases coral survival, so much so that when the 1998 El Nino killed 98 percent of the reef around Vabbinfaru in the Maldives, more than 80 percent of its corals survived, compared to just 2 percent elsewhere. Biorock can enable coral growth and regrowth, even in the presence of environmental stress, such as increasing water temperatures.

"Biorock reefs in Grand Turk survived the two worst hurricanes in the history of the Turks and Caicos Islands, which occurred three days apart and damaged or destroyed 80 percent of the buildings on the island," Goreau said. Not only did the biorock reef survive, but the sand built up around its structures.

Estimates for biorock reefs range from $20 to $1,290 per meter of shoreline depending on the size and shape of the reef grown, while other methods range from $60 to $155,000 per meter, or 3 to 120 times more expensive. A biorock shore protection reef was built in front of a beach resort that had washed away in the Maldives in 1997. Goreau says:

> The biorock reef was a linear structure parallel to the shore, 50 meters long, about 5 meters wide, and about 1.5 meters high, built on eroded reef bedrock. The structure cemented itself solidly to the limestone bedrock by mineral accretion. Waves were observed to slow down as they were refracted through the structure, dissipating energy by surface friction. Sand immediately began to accumulate on the shoreline and under and around the reef, and the beach grew back naturally and rapidly in a few years, and stabilized with no further erosion, even though the 2004 tsunami passed right over it. Corals growing on the biorock reef had 50 times (5,000 percent) higher coral survival than the adjacent natural coral reef after the 1998 coral bleaching event. For a decade after the bleaching event, this resort had the only healthy reef full of corals and fishes in front of their beach in the Maldives.

But then the hotel whose reef and beach were saved by the biorock project turned the power off, with the result that the corals, no longer protected from bleaching by the biorock process, suffered severe mortality in the 2016 bleaching event.

To continuously add corals and protect the reefs we still have, a continuous supply of low-current electricity is needed. This will be required for as long as the climate emergency and ocean acidification persists. The amount of power required is small, about one air conditioner's worth of electricity for Goreau's entire 400-meter-long Pulau Gangga beach restoration project in Indonesia, but global reef regeneration could be where another new technology might have a role to play—Ocean Mechanical Thermal Energy Conversion (OMTEC).

In the prototype OMTEC system under construction by Patrick McNulty off Florida's coast, the platform rests at the water's surface. It extends a vertical tunnel, 30 feet (9 m) in diameter, down 1,968 feet (600 m) nearly to the ocean floor. At one end of the platform, a large intake faces into the six-miles-per-hour Gulf Stream current, taking in warm water through a screen that keeps plastic, floating debris, and fish away from the intake. At the down-current end of the platform, the intake pipe narrows into a Venturi that drops pressure and increases flow speed. A turbine inside the Venturi captures that energy, but the pressure differential is enough to siphon cold water to the surface and send warm water down to the floor.

At the ocean floor, the water being drawn in is only about 6°C (43°F) compared to 22°C (72°F) at the surface. A 90-degree bend in the tunnel causes water to travel through a screen and into the tunnel and travel up toward the surface. Before it reaches the platform, however, it passes through a condenser coil filled with a liquid that boils into a gas at temperatures lower than the ocean surface. Because the pipe is insulated, the water coming from below is still cold enough to keep the fluid in its liquid state, but once the fluid is allowed to circulate up into the warmer waters above and turn into its gaseous form, it expands. That pressure is enough to drive a large electric generator on the platform. The gas is then directed back around the loop to cold water, where it liquifies and repeats the process.

Because cold water is released near the surface (heat-exchange dampers can adjust its exact temperature) and warm water is sent to the seafloor, if enough of these systems were built, they could significantly cool coastal waters, reducing the risk from hurricanes and large storms and cooling the atmosphere to buy time while we lower our emissions of greenhouse gases. There is enough thermal-mechanical power of this kind in the Gulf Stream to power human civilization 100 times over. A few thousand of these platforms, parked off the coast of Florida, would easily be enough to power the electric grid of North America. There are similarly favorable sites of predictably constant ocean circulation and temperature gradient in other regions.

A system like this could easily generate enough surplus energy, day and night through all seasons of the year, to restore and rebuild coral reefs everywhere, and continue to build the sandy beaches behind them, even as sea level rises.

More than 40 years ago, corrosion-protection systems employing pulsed electric currents were commonly used on offshore oil and gas facilities. They were known to cause limestone to form on the metal supports. As we move away from the era of fossil fuels, we have been leaving these rigs behind as artificial reefs on which coral can grow. Two of the most extensive arrays of oil rig reefs are in Indonesia, on the northwest coast of Bali, and at Gili Trawangan, on the northwest coast of Lombok Island. There are about 1,400 rigs in tropical Southeast Asia due for decommissioning and many more around the world.

At a pitch event in Tel Aviv in 2019, a start-up company called ECOncrete won the grand prize for their biologically supportive marine concrete using pozzolans, an ancient Roman method for building cement walls. ECOncrete's detailed geometry, as well as low acidity, speeds coral growth as much as 50 times (5,000 percent).

Two castable Reefcrete concrete mixes being tried out are made from Portland cement mixed with recycled ground and granulated blast-furnace slag or hemp fibers and recycled shell material. So far, the hemp and shell blends support significantly more live cover, with greater biodiversity, than the slag blend.

Gator Halpern, one of the founders of the Bahamas-based start-up Coral Vita, builds land-based coral farms near the reefs to be restored. These farms have large tanks that seawater passes through, and corals are grown under controlled conditions. Halpern's approach is to train his nursery corals to be more resilient to the ocean changes underway. He can crank up the heat or the acidity in the tanks and the corals can learn to survive and build resistance. Coral Vita also selectively breeds the individual corals that fare best and uses those genotypes to seed the next coral.

Corals grow in colonies of individual polyp animals. A single coral you see can actually be hundreds or thousands of these polyps. Halpern's team breaks the corals apart into single polyps and spaces them out. After that, they grow much more quickly and fuse back together into larger corals. At the nursery farm, they repeat this process over and over, breaking the corals apart, fusing them together, breaking them apart, fusing them, and eventually, after six to eight months, they plant those corals into the ocean.

If the natural growth rates of coral can be accelerated by a factor of up to 50 using industrial waste materials like abandoned oil rigs, then coupled with electro-stimulated biorock, we could conceivably grow coral reefs faster than they are dying off, and raise sandy coastlines faster than the ocean is rising.

Let's work back upstream to the destructive practices that lead to coral bleaching. Rene Castro Salazar, an assistant director-general at the UN Food and Agriculture Organization, says that two billion hectares (almost five billion acres) of land around the world have been degraded or desertified by misuse, overgrazing, deforestation, and other human activities. Nearly half of that is suitable to restore quickly, and the remainder can be recovered with time. We will need to do this to give the corals the clear water they need.

In December 2019, at a UN conference on desertification in New Delhi, 196 countries plus the European Union agreed to a declaration that each country would adopt measures needed to restore ecosystem health to unproductive land by 2030. A UN team used satellite imaging and other data to identify the 900 million hectares of degraded land that could be realistically restored today, much of that at a profit through increased food and water supplies, storm moderation, and other ecological benefits. "With political will and investment of about $300 billion, it is doable," Salazar said.

Returning the carbon to the soil and getting natural vegetation, pasture lands, and forests thriving again is a key to coral regeneration. Continuing to allow the soil to run off into the ocean or to add other contaminants to rivers spells disaster. We should be "using the least-cost options we have while waiting for the technologies in energy and transportation to mature and be fully available in the market," Salazar said. "It will stabilize the atmospheric changes, the fight against climate change, for 15–20 years. We very much need that."

We need to spend $300 billion, or $300 per hectare, to recover degraded estuaries and coastal wetlands, starting inland. That's the equivalent of the world's military spending every 60 days. That was the combined price tag for the cleanup after hurricanes Katrina and Harvey. What Salazar is saying is that it's also enough money to stop producing dead zones. It is enough to rescue many ocean life-forms now headed for extinction. It is enough to reverse the buildup of greenhouse gases and buy a few more years to prepare for global warming and sea-level rise that we cannot stop.

Restoration of a watershed can begin with reforesting the ridgelines and cleaning the streams that flow from them. At the lower end of the watershed lie estuaries, bays, and the open ocean, including mangrove forests, barrier islands, and coral reefs. This is not the Manhattan Project to build the atom bomb ($23 billion in today's dollars) or the Apollo Program to put a man on the moon ($152 billion), as its budget might suggest. Instead, we would be using about twice the cost of the first moon shot simply to keep millions of tons of acidifying carbon, fertilizer, and plastic trash from rushing into streams when it rains.

16

Reversing Climate Change

Ocean heat gain in 2019 was the largest on record. It will echo for centuries. The ocean is warming, acidifying, and losing oxygen, and sea level is rising. As a result, keystone species and ecosystems, such as warm-water coral reefs, seagrass meadows, and kelp forests, will face very high risks of damage and disappearance by the middle of this century, even if countries are able to reduce their carbon dioxide emissions, which most have not yet begun to do. As the ice is lost at the poles, sea level will rise around the world, flooding many coastal areas. More frequent and powerful storms, droughts, and floods will affect our ability to travel, grow food, and live in some places.

The loss of "blue carbon" in the form of coastal mangrove forests is especially tragic. Mangroves are super-sequestrators, piling up muddy sediments around their roots that are nutrient rich, allowing the trees and the myriad fish and crustaceans that live at the junction of freshwater and saltwater tides to reproduce and grow rapidly. They are also a ticking bomb for climate change when they are removed, because that carbon vault is unlocked and the sediments decay to send methane and carbon dioxide skyward.

Financial markets assume a relatively stable climate, but there is a threshold, or tipping point, beyond which risks can spike, damages can ruin insurance companies, and business bankruptcies can cascade. Already in India, the Middle East, and Australia, temperatures are pushing against the limits of human endurance. Rising temperatures are starting to limit the number of hours people can work, how much water is in the rivers, and whether food can be grown. In some places, even sleep is difficult.

For 15 years, Jeremy Grantham, cofounder of a Boston-based asset-management firm—credited with predicting the 2000 and 2008 downturns—has spoken out about the perils of the changing climate. He has concluded it's the number-one global risk. Grantham says the present system that manages the global economy is making the problem worse and has to change, quickly. "I used to talk about climate change, and my clients would roll their eyes and ask why I was wasting their time. But now everyone is at least talking about it," Grantham says. "The problem is, it's all talk. Last year more carbon-dioxide molecules went into the air than any single year in history. We must try harder."

If you say you are moving from an economy entirely based on burning carbon to something else, that is not a small change. That is a tectonic shift in the way modern society has been structured ever since we started burning coal, and then oil, and then natural gas. Most of the recommendations I make in this book have suggested we need better regulations and more of them. We need ocean cops on the beat. We need to ban more things.

And, while all of that may be true, it is a hard sell to many people, including a large number of voters and politicians. However, this century may yet offer us a softer path in the form of the digital currency revolution now underway. Since the digital age first reached central banks in the last quarter of the twentieth century, paper and coin money have taken a back seat to electronically stored and instantly transmitted strings of ones and zeros that make up the modern global economy. Each day quadrillions of dollars, euros, rubles, pesos, and yen are exchanged by keystroke. This revolution has now evolved into blockchains of digital ledgers that offer verifiability and chain-of-custody records, and one even more significant advantage. They offer the prospect that we may be able to de-externalize costs that harm society and the natural world.

When we cut down a tree to make paper or furniture, our ancient system of accounting counts that timber as an asset. As value is added through labor and technology, the wood appreciates and is assigned a higher value. We do not subtract from that value the work the tree had been doing that is now lost. We do not account for its role in moderating climate, freshening the air, or fostering biodiversity. But we could. The shift to distributed ledgers and the acceleration of computing power makes that kind of revaluation possible. It is already happening with experimental exchanges like Nori and Puro that calculate how carbon-sequestration value changes as a product or service is exchanged, ages, or recycles its components. Activity that benefits the climate conveys a higher value, while activity that reduces our security or damages the environment drops the value of the commodity.

The decision by many nations and cities to push for carbon neutrality by 2050 or earlier will bring opportunities to realign financial institutions to this new economic paradigm, where social and environmental costs are no longer externalized but are reflected in the price of anything exchanged. There will be opportunities for new jobs and better living conditions as a result. Still, it will involve massive new public and private investments and be painful for many workers in old-style carbon-burning professions.

The ocean offers opportunities to reduce the causes and consequences of climate change, globally and locally, if we adopt more ocean-inclusive strategies. Patricia Scotland, Secretary-General of the Commonwealth of Nations from 2015 to 2020, chose to lead by example. The Commonwealth has 53 member countries on six continents, approximately 32 percent of the world's total population (2.2 billion people)—more than 60 percent of whom are under age 29. The countries that make up the Commonwealth combine to account for the world's most extensive coastline. Though diverse, the Commonwealth has shared history, systems of law, and language. Scotland provided a central unifying belief: the need to address climate change and the future of the ocean collectively.

When she took office in 2015, the second Secretary-General from the Caribbean and the first woman to hold the post immediately convened panels of experts to tease out strategies to address the climate emergency. Exploring the range and diversity of available solutions, the Secretary-General identified problem and action areas; solutions that implement positive technical, behavioral, or policy change; and those that motivate people to act. Solutions, she recognized, had to be combined across these divides and disseminated in a practical way—one that raised awareness, facilitated collaboration, and was built for speed. From this understanding, she launched the *Commonwealth Blue Charter: Shared Values, Shared Ocean.*

"Sharing the ocean" is a curious way of framing the megatrend under way. In their study of blue acceleration, the Stockholm Resilience Center and the Royal Swedish Academy of Sciences identified 18 major competing claims that humans are imposing on the regenerative capacity of the ocean:

- Seafood
- Feed and nutraceuticals
- Hydrocarbons
- Minerals
- Desalinated water
- Ornamental resources (the aquarium trade)
- Genetic resources and products
- Scientific information
- Shipping
- Pipelines and cables
- Tourism and recreation
- Coastal land
- Renewable energy
- Geoengineering
- Waste disposal
- Conservation and regeneration
- Defensive boundaries
- Military navigation

Jean-Baptiste Jouffray and his coauthors in Stockholm write:

> Claiming marine resources and space is not new to humanity, but the extent, intensity, and diversity of today's aspirations are unprecedented. We describe this as the blue acceleration—a race among diverse and often competing interests for ocean food, material, and space.

If you run through that list of claims and try to arrange them into some kind of ranking system, you would get a widely divergent response, depending upon whom you asked. Someone concerned about national sovereignty might want defensive boundaries and military navigation. Someone concerned most about human welfare might choose food and desalinated water. An economist would like more hydrocarbons, minerals, shipping, pipelines, and cables. An environmentalist might prioritize waste disposal, coastal land, or renewable energy. My first choice would be for conservation and ecosystem regeneration.

Fortunately, we don't have to choose, but we may have to ration between the competing claims. Conservation measures such as turtle-safe beaches and restrictions against cutting mangroves can protect carbon-rich coastal ecosystems from disturbance and loss, and also make them less vulnerable to storms and flooding. The low-hanging fruit may be those strategies that give the most immediate benefits to the people engaged to implement them.

India has pledged to plant 13 million hectares of forest, much of it coastal mangroves, by 2020. Latin America is aiming at 20 million hectares, and African countries 100 million hectares, by 2030. China intends to plant an area of forest as large as Ireland every year, and Ireland itself has a trillion trees initiative.

But it isn't as simple as just grabbing seeds and saplings and sticking them in the ground. Non-native plantations can cause problems for biodiversity or local livelihoods—or both. Monoculture forests are unnatural, become diseased, and attract pests. What counts are survival rates, not just for the saplings, but also for the ecosystems and local cultures of whose broader context those trees are a part. For this reason, any serious effort at ecosystem restoration also involves a system of caregiver training, follow-up, mixed-age and mixed-species forests, permaculture, and cultural integration into the health of the forest, or, put another way, forest integration into the health of the human community.

Reforestation has been the focus of the Ecosystem Restoration Camps begun by John D. Liu and its partners at the Global Ecovillage Network and the Permaculture Research Institute. Liu's concept, now running in several countries and supported by thousands of volunteers, followers, and advisors, is to convene "camps" where young (and older) campers could gather to build earthworks, plant trees, and restore damaged landscapes. Using the model of the Loess Plateau in China as an inspiration, these camps are demonstrating that ecosystem regeneration works and can be fun too.

Restoring and enhancing coastal vegetation better adapts the coasts to warmer summers and colder winters, extreme storms, and erosion while contributing to food security and biodiversity. Nearly 50 percent of coastal wetlands have been lost over the last 100 years as a result of the combined effects of localized human pressures, sea-level rise, warming, and extreme climate events. Vegetated coastal ecosystems are essential carbon stores; their loss is responsible for as much as one and a half billion tons of carbon dioxide per year—about 4 percent of total greenhouse pollution.

Because of the warming ocean, distribution ranges of seagrass meadows, mangroves, and kelp forests have been expanding at high latitudes and contracting at low latitudes since the late 1970s and, in some areas, have been devastated by heat waves. Large-scale mortality is related to warming, but most often it's due to coastal and port development and eutrophication of inland waterways. Since the 1960s, the decline has been partially offset by migration of mangrove forests into subtropical salt marshes as low coastal areas gradually become part of the sea.

These newly forming mangrove forests seem to be keeping pace with sea-level rise in many parts of the world. If we are willing to deconstruct barriers like roads and seawalls that prevent a landward shift of marshes and mangroves (engineers call this "coastal squeeze"), wetlands and mangroves can grow land vertically at rates equal to or higher than current mean sea-level rise.

We can assist this process by restoring former wetlands drained for human settlements and by intensively planting newly forming wetlands with carbon-storing mangroves. By increasing photosynthetic production, new vegetation can speed carbon drawdown at a scale that is locally and globally significant. These so-called "blue carbon" coastal ecosystems build biodiversity, clean polluted waters, and return many natural services we were losing without them. We could be taking millions of tons of carbon from the atmosphere immediately and continually if we had a massive effort to restore seagrass meadows and to farm seaweed.

One organization that places particular importance on coastal mangroves is TreeSisters, initially in the UK but now beginning in the US and Australia also. Its focus is primarily on regions where urgent forest restoration or conservation can prevent further damage to the last frontiers of remaining ancient forests, and on regions where social development goals can be achieved while planting trees.

TreeSisters is made of regional groups, called "groves," that provide training and support to local women's efforts. Why women? First, because in the overdeveloped world, women make over 85 percent of consumer choices, and many of those choices drive deforestation. Second, as groups such as Project Drawdown have identified, empowerment and education of women and girls is the number one most effective strategy to reverse climate change.

Eden Reforestation Projects, one of the TreeSisters' partners, employs more than 1,000 people to plant trees, with 225 million new mangrove trees planted since 2006. Some Malagasy planters were enslaved to local fish barons because they owed money for their fishing equipment. Tree-planting income, raised by TreeSisters through social media crowdsource campaigns, has enabled them to repay their debts and escape their bondage. In 2019, Eden launched new mangrove projects in Mozambique and West Papua and in 2020 will be planting food forests in the Amazon with and through AquaVerde and the Ashaninka tribe. TreeSisters is in Madagascar and Nepal (with Eden), Brazil and Northern India (with WeForest), Cameroon and Kenya (with International Tree Foundation), and Southern India (with Project GreenHands). It is exploring more projects in Colombia, the Democratic Republic of the Congo (one of the most violent nations for women), and Uganda. Project GreenHands currently seeks support to plant 114 million trees in the shortest span possible in coastal Tamil Nadu.

Pollution reduction in coastal waters removes contaminants and excess nutrients that impair ecosystem functions and, in that way, can help grow the capacity to withdraw carbon from the atmosphere and the ocean. Reduced pollution from shipping can also, to a limited degree, address the causes of climate change by keeping fossil fuels out of the

THE CLIMATE EMERGENCY

In 2018, the Club of Rome became one of the leading international organizations to issue a declaration of climate emergency. To arrest the rate of climate change acceleration, it recommended 10 priority actions:

1. **Halt fossil fuel expansion and fossil fuel subsidies by 2020.** No new investments in coal, oil, and gas exploration and development after 2020 and a phaseout of the existing fossil fuel industry by 2050. A phaseout of fossil fuel subsidies by 2020.

2. **Triple annual investments in renewable energy,** energy efficiency, and low-carbon technologies for high-emitting sectors before 2025. Give priority to developing countries to avoid lock-in with the carbon economy.

3. **Put a price on carbon to reflect the true cost** of fossil fuel use and embedded carbon by 2020. Introduce carbon floor prices. Tax embedded carbon through targeted consumption taxes. Direct tax revenues to research, develop, and innovate low-carbon solutions.

4. **Replace GDP growth as the main objective for societal progress** and adopt new indicators that accurately measure welfare and well-being rather than production growth.

5. **Improve refrigerant management by 2020.** Adopt ambitious standards and policy to control leakages of refrigerants from existing appliances through better management practices and recovery, recycling, and destruction of refrigerants at the end of life.

atmosphere. This is the type of initiative launched by Patricia Scotland's Commonwealth Secretariat through its Clean Ocean Alliance.

Boyan Slat is the young Dutch engineer who founded The Ocean Cleanup, a nonprofit that designs technologies to remove plastic from the ocean. In 2019, he collected the first two shipping containers of trash from the Great Pacific Garbage Patch by deploying surface traps along the path of ocean currents. As plastic wastes drove themselves toward his devices, they were gathered and moved to shore for recycling. The first models are capturing particulates down to one millimeter in size.

After learning that 1 percent of rivers are responsible for 80 percent of plastic emissions, Ocean Cleanup developed the Interceptor, a solar-powered device that, when placed in river mouths, catches the

6. **Encourage exponential technology development by 2020.** Create an international task force to explore alignment of exponential technologies and business models with the Paris Agreement to promote technology disruption in sectors where carbon emissions have been difficult to eliminate.

7. **Ensure greater materials efficiency and circularity by 2025.** Significantly reduce the impact of basic materials—e.g., steel, cement, aluminum, and plastics—from almost 20 percent of global carbon emissions today by the early introduction of innovation, materials substitution, energy efficiency, renewable energy supply, and circular material flows.

8. **Accelerate regenerative land use policies and adaptation.** Triple annual investments in large-scale REDD+ reforestation and estuarine marshland initiatives in developing countries. Compensate farmers for building carbon in the soils and promote forestry sequestration. Support efforts to restore degraded lands. Implement adaptive risk management procedures in every state, industry, city, or community.

9. **Ensure that population growth is kept under control** by giving priority to education and health services for girls and women. Promote reproductive health and rights, including family planning programs.

10. **Provide for a just transition in all affected communities.** Establish funding and retraining programs for displaced workers and communities. Provide assistance in the diversification of higher carbon industries to lower carbon production. Call upon the top 10 percent earners of the world to cut their GHG emissions by half by 2030.

plastic before it reaches the oceans. The first Interceptors are now active in Indonesia and Malaysia and should be in all the heaviest-polluting rivers soon.

That said, it is no excuse for people to continue to consume and dispose of single-use plastic, much of which is nonrecyclable. People must refuse, reuse, and return all the plastics they are handed in stores.

Community-based adaptation and risk reduction policies might include relocating populations back from vulnerable coasts and building storm-resistant structures to protect ports. These can also include steps we all know are needed to modify society to reverse climate change, including eating less meat and wild fish; using energy more efficiently; substituting natural, biodegradable fabrics and biopolymers for fossil plastic and synthetics; switching to shared transportation powered by renewable energy sources; and growing organic gardens.

One of the more critical steps most climate initiatives have overlooked is letting whales live. If whales were allowed to return to their pre-whaling number of four to five million—from slightly more than 1.3 million today—it could add significantly to the amount of phytoplankton in the oceans and to the carbon they capture each year. At a minimum, even a 1 percent increase in phytoplankton productivity thanks to whale activity would capture hundreds of millions of tons of additional CO_2 a year, equivalent to the sudden appearance of two billion mature trees.

Despite the drastic reduction in commercial whaling, whales still face significant life-threatening hazards, including ship strikes, entanglement in fishing nets, waterborne plastic waste, and noise pollution. While some species of whales are recovering—slowly—many, like the blue whale, are dangerously close to extinction.

A potential model solution can be seen in the United Nations program to end deforestation and desertification (REDD). Recognizing that deforestation accounts for 17 percent of carbon emissions, REDD provides incentives for countries to preserve their forests as a means of keeping CO_2 out of the atmosphere. Similarly, we could create financial mechanisms to promote the restoration of the world's whale populations, such as subsidies or other compensation to protectors and responsible marine industries. Whales are commonly found in the waters around low-income and fragile states, countries that may be

unable to pay for the needed protection. Enhancing the protection of whales from human-made dangers would deliver benefits to ourselves, the planet, and, of course, the whales. This "earth-tech" approach to carbon sequestration also avoids the risk of unanticipated harm from suggested, untested, high-tech fixes. Nature has had millions of years to perfect her whale-based carbon sink technology. All we need to do is let the whales do their magic.

In saving the seas, it may sometimes be necessary to set priorities and help ocean populations and ecosystems most at risk. For this reason, one of the essential strategies must be to protect and expand marine reserves. This is a simple and straightforward tool we have available at almost no cost. The cost is relocating or retraining human populations that are impacting reserve areas and providing remote sensing, patrol boats, courts, and legal structures to enforce the borders of protected waters.

Many forms of renewable energies, such as tidal, thermal, current, solar photovoltaic, and wind-generated electricity, are carbon-neutral and will help get the greenhouse gas footprint closer to zero. However, newer green technologies promise to go beyond zero emissions and decarbonize both the atmosphere and the ocean, reversing both climate change and ocean acidification.

If we want to address the causes of climate change, we have to go a little deeper. We need to look closely at each of our individual lifestyle choices and identify areas of improvement. The ocean we leave for our children depends upon it.

Conclusion

Given what we have seen about the blue acceleration, starting centuries ago but clicking into high gear in the twenty-first century, it will be no small challenge to bend our present trajectory to find a satisfactory outcome for all the competing claims on the ocean. With rapid climate change unfolding even as we read this, highly unpredictable events are likely to continue throughout our lifetimes.

Imagine for a moment . . .

Xen becomes an architect. By the time he goes to university, his tutors are no longer emphasizing architecture by the examples of Le Corbusier, Wright, Gropius, and I. M. Pei, but instead looking to the insights of Bill Mollison, William McDonough, Ellen MacArthur, and Ianto Evans. We see a design process that offers a positive, regenerative future for the planet. Buildings and streets do not embody exhaustible energy and nonrenewable materials, but, mimicking natural processes, they absorb and hold CO_2, moderate their own temperatures, and contribute a favorable microbiome to their surrounds. Urban neighborhoods produce their own energy, food, commerce, and conviviality. If you follow the path of products that pass through Xen's life, they no longer move in one direction from mine to landfill, polluting streams and oceans along the way. Instead, they travel from former product to future product in a spiral of improvement and upcycling.

Other changes may transform Xen's world in ways we can only barely imagine. Xen may find himself designing floating city blocks recycled from the hulls of cargo transports, oil tankers, and aircraft carriers; autonomous maritime transportation vessels propelled by wind; underwater hydroponic farms growing food safely away from the unpredictable storms on the land; or deep-ocean data farms using thermocline cooling to speed their microprocessors.

Xen's chosen profession, like most at mid-century, will produce a broad spectrum of positive effects—enhancing environmental health, social well-being, and economic vitality. His and his coworkers' design process will be built on a dialogue with nature—an interaction with sunlight, wind, water, and nutrient cycles, seen and unseen, close and far. By integrating their dialogue into not only architecture but also into transforming all of human industry, new ecosystems of relationships will form and innovation will accelerate. The buildings that Xen designs will join into loops of material and biological flows for their heating and cooling, green roofs, water-gathering balconies, and stormwater filters. The buildings themselves will make oxygen, fix nitrogen, distill water, and provide habitat for diverse species, seen and unseen. Even as they rest on movable or immovable foundations, they will restore landscapes, expand biodiversity, and change the atmosphere miles above them. The neighborhoods and districts Xen plans may shift how they appear and function with the seasons but will become beautiful, safe, comfortable, and deeply satisfying places to live, work, learn, and participate in the culture of a planet-nourishing community.

By the time Xen reaches the midpoint of his career, he will have witnessed many changes that might astonish us. World population will have bent the curve and be coming back down to a level that can be sustained ecologically. Food will be produced in many new and different ways, and none of those will harm the world. Wild animals will be expanding their range and numbers, on land and in the sea.

Humans will have passed from carefree juveniles to grown-ups—taking responsibility for their species and its role as an adult on the Earth. All over the world, people are being trained to be Emergency Planetary Technicians and sent far and wide to arrest the spread of deserts, repair coral reefs, and restore wetlands.

Of course, Xen's generation could not entirely stop climate change. What had begun centuries earlier will take centuries to unwind. Three degrees was inevitable by the time Xen left home for university, but by then, the world had enough of a foretaste of the consequences that it had begun to act. Fossil fuels stopped being taken up from the ground and burned. Plastics became bioplastics that unzipped into harmless elements on contact with seawater. Ships, cars, and trains

ran on sunlight. Forests expanded to cover what had been starting to be deserts. Soils grew deep and black. Birds and bears returned. Glaciers reformed at high latitudes and on mountaintops.

Perhaps the single greatest challenge Xen's generation had to face was the growing demand for the health value of seafoods that was underway when he was eight years old. His generation succeeded in saving the marine environment and the turtles and whales because they took the steps required to make laws protecting marine populations and that punished bad actors in the fishing, shipping, mining, and other ocean industries. They showed the demand for healthy seafoods could be met in a way that was healthy for everyone.

The ocean is not merely our birthplace. It is our mother. It evaporates saltwater in the warming rays of the sun and sends sweet water for us to drink. Our planet's spin and the light arriving from the sun combine to generate waves and wind, and that wind blows across continents to freshen our air, turn our wind generators, and dry our clothes. Phytoplankton gardens make more oxygen than all the forests on land. Great whales fertilize these blooms, and gravity carries these plants and animals deep when they die, balancing the carbon in Earth's atmosphere. That, in turn, regulates the climate we, and all of creation, need to survive and thrive.

The health and well-being of all of us depends on keeping the ocean happy.

REFERENCES

Abulafia, D., *The Boundless Sea: A Human History of the Oceans* (2019).

Aggarwal, M., "India has brought 9.8 million hectares of degraded land under restoration since 2011," *Mongabay* (4 Sep 2018).

Anon., "Operation Crossroads," Twice Risen Sun (2017).

Anon., "Seal level rise accelerating on US coastline, scientists warn," *The Guardian* (3 Feb 2020).

Barkham, P., "Can planting billions of trees save the planet?" *The Guardian* (19 Jun 2019).

Bendell, J., "Deep adaptation: A map for navigating climate tragedy," *IFLAS Occasional Paper 2* (2018).

Bergmann, M. et al. *Marine Anthropogenic Litter* (2019).

Bernaldo de Quirós, Y. et al., "Advances in research on the impacts of anti-submarine sonar on beaked whales," *Proceedings of the Royal Society B* 286:1895 (2019).

Biello, D., "Can Geoengineering Save the World from Global Warming?" *Scientific American* 25 (Feb 2011).

Brandin, A. S., Environment, "2019: The year of climate consciousness," *DW News* 27 (2019).

Caine, A., *Marine Biology for the Non-Biologist* (2015).

Caldeira, K., and M. E. Wickett, "Ocean model predictions of chemistry changes from carbon dioxide emissions to the atmosphere and ocean," *Journal of Geophysical Research: Oceans* 110:C9 (2005).

Chami, R. et al., "Nature's Solution to Climate Change," *Finance & Development* 56:4 (2019).

Cheung, W. et al., "Signature of ocean warming in global fisheries catch," *Nature* 497.7449 (2013).

China Daily, "WMO: 2019 was 2nd hottest year ever, more extreme weather" (16 Jan 2020).

Coates, K. J., "Warming waters hurt Zanzibar's seaweed," *Christian Science Monitor* (21 May 2018).

Corbett, J., "'Scale of This Failure Has No Precedent': Scientists Say Hot Ocean 'Blob' Killed One Million Seabirds," Common Dreams (Jan 2020).

Costanza, R. et al., "Quality of life: An approach integrating opportunities, human needs, and subjective well-being," *Ecological Economics* 61:2–3 (2007).

CRC Research, *Let's all Breathe Easier* (2019).

Daly, H. E., and J. Farley, *Ecological Economics: Principles and Applications,* 1st ed. (2004).

Dennis, H. D. et al., "Reefcrete: Reducing the environmental footprint of concretes for eco-engineering marine structures," *Ecological Engineering* 120 (2018).

Dunn, N., "What climate change effects will a Greenland shark born today experience?" *Grantham Institute—Climate Change and the Environment* (2019).

Economist Group, "Is deep-sea mining part of the blue economy?" *World Ocean Initiative* (2019).

Englander, J., *High Tide on Main Street: Rising Sea Level and the Coming Coastal Crisis,* ed. 2.2 (2013).

Farley, J., and R. Costanza, "Envisioning shared goals for humanity: A detailed, shared vision of a sustainable and desirable USA in 2100," *Ecological Economics* 43:2 (2002).

Farley, J. et al., "Synthesis: The quality of life and the distribution of wealth and resources," *Understanding and Solving Environmental Problems in the 21st Century: Toward a New, Integrated Hard Problem Science* (2002).

Feely, R. A., S. C. Doney, and S. R. Cooley, "Ocean acidification: Present conditions and future changes in a high-CO world," *Oceanography* 22:4 (2009).

Fehr, E. and K. M. Schmidt, "The economics of fairness, reciprocity and altruism: Experimental evidence and new theories," *Handbook on the Economics of Giving, Reciprocity and Altruism*, Vol. 1 (2006).

Fisher, M., "40 maps that explain the Middle East," Vox (26 Mar 2015).

Flannery, T., *Sunlight and Seaweed: An Argument for How to Feed, Power and Clean Up the World* (2017).

Gaille, B., "41 Fishing Industry Statistics and Trends," BrandonGaille Small Business and Marketing Advice (2020).

Gaille, B., "43 Seafood Industry Statistics, Trends & Analysis," BrandonGaille Small Business and Marketing Advice (2018).

Gattuso, J. P. et al., "Contrasting futures for ocean and society from different anthropogenic CO_2 emissions scenarios," *Science* 349:6243 (2015).

Giggs, R., "Human drugs are polluting the water—and animals are swimming in it," *The Atlantic* (May 2019).

Golden, J. S. et al., "Making sure the blue economy is green," *Nature Ecology & Evolution* 1:2 (2017).

Goodell, J., "Goodbye, Miami," *Rolling Stone* 20 (2013).

Gooley, T., *How to Read Water* (2019).

Goreau, T. J. F., and P. Prong, "Biorock electric reefs grow back severely eroded beaches in months," *Journal of Marine Science and Engineering* 5:4 (2017).

Gosztola, K., "UN Report: Marine Heat Waves Hurtling Planet Toward Tipping Point," Shadowproof (25 Sep 2019).

Griggs, R., "Pharmaceutical Pollution Hurts Wild Animals," *The Atlantic* (May 2019).

Hall, C. M., "Trends in ocean and coastal tourism: the end of the last frontier?" *Ocean & Coastal Management* 44:9–10 (2001).

Hamilton, C., "Ethical anxieties about geoengineering," *Ethics and Emerging Technologies* (2014).

Harvard Health Publishing, Harvard Health Letter, "Drugs in the water" (Jun 2011).

Harvell, C. D. et al, "Emerging marine diseases—climate links and anthropogenic factors," *Science* 285:5433 (1999).

Hatermann, B., *Marine Science for Kids: Exploring and Protecting Our Watery World* (2017).

Howarth, R. W., "Methane emissions and climatic warming risk from hydraulic fracturing and shale gas development: implications for policy," *Energy and Emission Control Technologies* 3 (2015).

Huang, R. J. et al., "High secondary aerosol contribution to particulate pollution during haze events in China," *Nature* 514:7521 (2014).

International Maritime Organization, "Marine geoengineering including ocean fertilization to be regulated under amendments to international treaty, Briefing 45," *35th Consultative Meeting of Contracting Parties to the Convention on the Prevention of Marine Pollution by Dumping of Wastes and Other Matter, 1972 (London Convention)* (2013).

Johnson, A., "Our oceans brim with climate solutions. We need a Blue New Deal," *Washington Post* (10 Dec 2019).

Jouffrey, et al., *The Blue Acceleration: The Trajectory of Human Expansion, One Earth* (2020).

Kabisch, Nadja et al., "Nature-based solutions to climate change mitigation and adaptation in urban areas: perspectives on indicators, knowledge gaps, barriers, and opportunities for action," *Ecology and Society* 21:2 (2016).

Kituyi, M., and P. Thomson, "Nearly 90% of fish stocks are in the red—fisheries subsidies must stop," *World Economic Forum* (13 Jul 2018).

Kubiszewski, I., J. Farley, and R. Costanza, "The production and allocation of information as a good that is enhanced with increased use," *Ecological Economics* 69:6 (2010).

Laville, S., "Thousands of ships could dump pollutants at sea," *The Guardian* (29 Oct 2018).

Lewis, S., "Scientists alarmed to discover warm water at 'vital point' beneath Antarctica's 'doomsday glacier,'" *CBS News* (1 Feb 2020).

Masters, J., "The Thawing Arctic: Risks and Opportunities," Council on Foreign Relations, (16 Dec 2013).

Max-Neef, M., "Development and human needs," *Real-Life Economics: Understanding Wealth Creation* (1992).

MENA, *Strengthening of National Capacities for the Implementation of the Nagoya Protocol on Access to Genetic Resources and the Fair and Equitable Sharing of Benefits Arising from Their Utilization to the Convention on Biological Diversity* (2014).

Mengel, M. et al., "Committed sea-level rise under the Paris Agreement and the legacy of delayed mitigation action," *Nature Communications* 9:601 (2018).

Mesley, P., "Bikini Atoll, submerged nuclear aftermath," *Advanced Diver Magazine* (July 2014).

Moir, R., "The narwhals' tale of rising seawater and sinking ocean," *Clam Chowdah* (2018).

Osborne, H., "There's a chance the North Atlantic current will shut down temporarily in the next 100 years," *Newsweek* (31 Dec 2019).

Pachauri, R. K. et al., *Climate Change 2014: Synthesis Report. Contribution of Working Groups I, II and III to the Fifth Assessment Report of the Intergovernmental Panel on Climate Change*, IPCC (2014).

Pauli, G. A., *The Blue Economy: 10 Years, 100 Innovations, 100 Million Jobs*, Paradigm Publications (2010).

Pauly, D., R. Watson, and J. Alder, "Global trends in world fisheries: impacts on marine ecosystems and food security," *Philosophical Transactions of the Royal Society B: Biological Sciences* 360.1453 (2005).

Pearce, F., *With Speed and Violence: Why Scientists Fear Tipping Points in Climate Change* (2007).

Pollard, S., *Puget Sound Whales for Sale* (2014).

Radio Ecoshock, "Ocean Heat Warning!" (12 Feb 2020).

Reilly, P., "Cruise ship damages pristine coral reef. How big an impact do cruises have on the environment?" *The Christian Science Monitor* (15 March 2017).

Roberts, C., *The Ocean of Life* (2015).

Rosane, O., "World cities that could be underwater as oceans rise," Ecowatch (5 Oct 2018).

Rust, S., "How the U.S. betrayed the Marshall Islands," *Los Angeles Times* (10 Nov 2019).

Samuelson, A., "New Earth mission will track rising oceans into 2030," NASA Jet Propulsion Laboratory (Nov 2019).

Sofiev, M. et al., "Cleaner fuels for ships provide public health benefits with climate tradeoffs," *Nature Communications* 9:406 (2018).

Stahl, B., *Save the Ocean (Save the Earth Book 1)* (2019).

Stephenson, F. R., L. V. Morrison, and C. Y. Hohenkerk, "Measurement of the Earth's rotation: 720 BC to AD 2015," *Proceedings of the Royal Society A* (2016).

Stokstad, E., "Fishing fleets have doubled since 1950 but they're having a harder time catching fish," *Science News* (27 May 2019).

Stumpf, M. et al., "Polar drug residues in sewage and natural waters in the state of Rio de Janeiro, Brazil," *Science of the Total Environment* 225:1–2 (1999).

Suzy's Animals of the World Blog, "The Octopus" (2019).

Sweeting, J. E. N., and S. L. Wayne, "A shifting tide: environmental challenges and cruise industry responses," *Cruise Ship Tourism* 30:327 (2006).

Syracuse University, "After all of this effort only 5 percent of the seafloor has been mapped," BIO 100 (2019).

Turney, C. S. et al., "Early Last Interglacial ocean warming drove substantial ice mass loss from Antarctica," *Proceedings of the National Academy of Sciences* (2020).

VanderMeer, J., "An op-ed from the future: It's 2071, and we have bioengineered our own extinction; the micro- and macro-organisms that saved humanity from our climate crisis are now changing us—and might destroy us," *The New York Times* (9 Dec 2019).

Vergara, W. et al., "The climate and development challenge for Latin America and the Caribbean: Options for climate-resilient, low-carbon development," Inter-American Development Bank (2013).

Walker, R., "'Deep Adaptation' is not based on science—climate change draft paper with unfounded claims of near future human extinction," *Science* 20 (19 Nov 2019).

World Bank, *How Does Port Efficiency Affect Maritime Transport Costs?* (2019).

World Oceans Forum, "Key takeaways from the *United Nations Special Report on Oceans and the Cryosphere*" (2019).

World Wildlife Fund, "Polar bear assessment brings good and troubling news" (2019).

Wu, J., "Major UN climate report says rapid ocean warming is causing 'heat waves,' threatening fishing industry," *CNBC* (25 Sep 2019).

Xia, R., "California coastal waters rising in acidity at alarming rate, study finds," *Los Angeles Times* (16 Dec 2019).

Yale 360, "As Oceans Warm, the World's Kelp Forests Begin to Disappear" (2019).

Yin, J. et al., "Different magnitudes of projected subsurface ocean warming around Greenland and Antarctica" *Nature Geoscience* 4:8 (2011).

ALBERT BATES is a former attorney, paramedic, aid worker, communard, natural builder, educator and the author of eighteen books on climate, history, and ecology, including *Taming Plastic: Stop the Pollution* (2020), *Transforming Plastic: From Pollution to Evolution* (2019), *BURN: Using Fire to Cool the Earth* (2019), *The Paris Agreement* (2015), *The Biochar Solution: Carbon Farming and Climate Change* (2010), and *The Post-Petroleum Survival Guide and Cookbook* (2006). His book *Climate in Crisis* (1990) was among the first to call attention to the potential for a runaway greenhouse effect in the twenty-first century. Since 1984 he has been the director of the Global Village Institute for Appropriate Technology (gvix.org), a nonprofit scientific research, development, and demonstration organization with projects on six continents, an ambassador for the Global Ecovillage Network, and an advisor to many organizations, foundations, and governments now applying regenerative design to reverse climate change.

GROUNDSWELL BOOKS
SOLUTIONS FOR A SUSTAINABLE WORLD

For more books that inspire readers to create a healthy,
sustainable planet for future generations, visit
BookPubCo.com

PLANET IN CRISIS SERIES

Addresses the urgent challenges of climate
change by focusing on specific issues,
identifying their impact, and illustrating
creative solutions that can make a difference.

Transforming Plastic
From Pollution to Evolution
Albert Bates

978-1-57067-371-9 • $9.95
128 pages • 6 x 9 paper • black & white

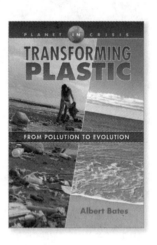

PLANETARY SOLUTIONS SERIES

Inspiring young people to understand,
challenge, and solve the environmental
problems that put the Earth at risk.

Taming Plastic
Stop the Pollution
Albert Bates

978-1-939053-24-4 • $14.95
48 pages • 8 x 9½ paper • full color

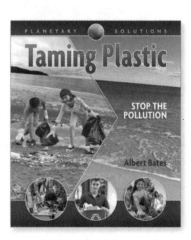

Purchase these titles from your favorite book source or buy them directly from:
Book Publishing Company • PO Box 99 • Summertown, TN 38483 • 1-888-260-8458
Free shipping and handling on all orders